AF735191

Cognitive Security for Industrial IoT

Industry 4.0, marked by the integration of cyber-physical systems, automation, and smart manufacturing, has greatly enhanced the intricacy and interconnectedness of industrial operations. Industrial Internet of Things (IIoT) systems allow for smooth communication and collaboration among machines, devices, and sensors, enabling real-time monitoring, data collection, and decision-making. As a result, this increased connectivity leaves IIoT networks vulnerable to a wider array of cyber threats, underscoring the need for more advanced security protocols.

Cognitive Security for Industrial IoT: AI-Driven Approaches to Industry 4.0 examines cognitive security principles for IIoT environments, emphasizing artificial intelligence (AI) and machine learning (ML) in security measures. This book provides practical case studies and real-world examples while addressing the unique security challenges inherent in Industry 4.0 technologies. Key focus areas include predictive maintenance, proactive threat mitigation, behavioral analytics, emerging trends, and best practices for implementing cognitive security in industrial environments.

This comprehensive resource explores cutting-edge trends and future prospects in IIoT security, including quantum-resilient technologies and decentralized security frameworks. Readers will find step-by-step guidelines and proven best practices for implementing cognitive security in industrial settings. This practical approach equips both practitioners and organizations to effectively combat evolving cyber threats, making it an invaluable resource for staying ahead in cybersecurity's rapidly changing landscape.

Internet of Everything (IoE): Security and Privacy Paradigm
Series Editors: Vijender Kumar Solanki, Raghvendra Kumar, and Le Hoang Son

IOT

Security and Trust Issues in Internet of Things
Blockchain to the Rescue
Edited by Sudhir Kumar Sharma, Bharat Bhushan, and Bhuvan Unhelkar

Internet of Medical Things
Paradigm of Wearable Devices
Edited by Manuel N. Cardona, Vijender Kumar Solanki, and Cecilia García Cena

Integration of WSNs into Internet of Things
A Security Perspective
Edited by Sudhir Kumar Sharma, Bharat Bhushan, Raghvendra Kumar, Aditya Khamparia, and Narayan C. Debnath

IoT Applications, Security Threats, and Countermeasures
Edited by Padmalaya Nayak, Niranjan Ray, and P. Ravichandran

Multimodal Biometric Systems
Security and Applications
Edited by Rashmi Gupta and Manju Khari

Smart Healthcare Monitoring Using IoT with 5G
Challenges, Directions, and Future Predictions
Edited by Meenu Gupta, Gopal Chaudhary, Victor Hugo C. de Albuquerque

Information Security Handbook
Edited by Noor Zaman, Khalid Hussain, Mamoona Humayun, Azween Bin Abdullah, and João Manuel R.S. Tavares

Cognitive Security for Industrial IoT
AI-Driven Approaches to Industry 4.0
Edited by Sudan Jha, Vijender Kumar Solanki and Maurice E. Dawson

For more information about this series, please visit: www.routledge.com/Internet-of-Everything-IoE/book-series/CRCIOESPP

Cognitive Security for Industrial IoT

AI-Driven Approaches to Industry 4.0

Edited by
Sudan Jha, Vijender Kumar Solanki and
Maurice E. Dawson

CRC Press
Taylor & Francis Group
Boca Raton London New York

CRC Press is an imprint of the
Taylor & Francis Group, an **informa** business

Designed cover image: Shutterstock

First edition published 2026
by CRC Press
2385 NW Executive Center Drive, Suite 320, Boca Raton FL 33431

and by CRC Press
4 Park Square, Milton Park, Abingdon, Oxon, OX14 4RN

CRC Press is an imprint of Taylor & Francis Group, LLC

© 2026 selection and editorial matter, Sudan Jha, Vijender Kumar Solanki and Maurice E. Dawson; individual chapters, the contributors

Reasonable efforts have been made to publish reliable data and information, but the author and publisher cannot assume responsibility for the validity of all materials or the consequences of their use. The authors and publishers have attempted to trace the copyright holders of all material reproduced in this publication and apologize to copyright holders if permission to publish in this form has not been obtained. If any copyright material has not been acknowledged please write and let us know so we may rectify in any future reprint.

Except as permitted under U.S. Copyright Law, no part of this book may be reprinted, reproduced, transmitted, or utilized in any form by any electronic, mechanical, or other means, now known or hereafter invented, including photocopying, microfilming, and recording, or in any information storage or retrieval system, without written permission from the publishers.

For permission to photocopy or use material electronically from this work, access www.copyright.com or contact the Copyright Clearance Center, Inc. (CCC), 222 Rosewood Drive, Danvers, MA 01923, 978-750-8400. For works that are not available on CCC please contact mpkbookspermissions@tandf.co.uk

For Product Safety Concerns and Information please contact our EU representative GPSR@taylorandfrancis.com. Taylor & Francis Verlag GmbH, Kaufingerstraße 24, 80331 München, Germany.

Trademark notice: Product or corporate names may be trademarks or registered trademarks and are used only for identification and explanation without intent to infringe.

ISBN: 978-1-041-00358-8 (hbk)
ISBN: 978-1-041-00426-4 (pbk)
ISBN: 978-1-003-60977-3 (ebk)

DOI: 10.1201/9781003609773

Typeset in Times
by Newgen Publishing UK

Contents

Preface

The convergence of operational technology (OT) with information technology (IT) through the Industrial Internet of Things (IIoT) is reshaping global industries. This integration, while driving remarkable gains in productivity and innovation, introduces profound cybersecurity vulnerabilities. Critical infrastructure, from power grids to manufacturing floors, is now a potential target for cyberattacks with real-world physical consequences. This book, *Cognitive Security for Industrial IoT: AI-Driven Approaches to Industrial 4.0*, arises from the urgent need to address these vulnerabilities with more intelligent and adaptive solutions.

Our aim in curating this work is to provide a comprehensive overview of how cognitive computing, Artificial Intelligence (AI), and Machine Learning (ML) are revolutionizing security in industrial environments. The book is designed to guide readers from the fundamental concepts of IIoT architecture and AI fundamentals to the advanced application of these technologies for robust threat detection, anomaly identification, and proactive system defense. The diverse perspectives included herein underscore a unified theme: that security must evolve from a static safeguard into a dynamic, learning, and predictive component of the IIoT ecosystem itself.

We extend our deepest gratitude to the cohort of international experts and researchers who have contributed their invaluable insights and groundbreaking work to this volume. Their chapters represent the forefront of cognitive security research, offering both theoretical depth and practical wisdom. We also thank the editorial and production teams for their unwavering support and dedication to this project.

This book is intended for a broad audience, including academic researchers, industry practitioners, cybersecurity specialists, and advanced students. It is our hope that the knowledge contained in these pages will not only inform, but also inspire further innovation, fostering the development of resilient and secure industrial systems for a smarter, safer future.

The Editors

About the Editors

Sudan Jha is a Professor in the Department of Computer Science and Engineering at Kathmandu University, Nepal, with over 23 years of teaching, research, and industrial experience as a senior Institute of Electrical and Electronics Engineers (IEEE) member. He leads research at the IoT R&D lab at Kathmandu University and previously served as Technical Director at Nepal Television, Principal at Nepal College of IT, and Individual Consultant at Nepal Telecom Authority. Prof. Jha has published 80+ research papers and book chapters, serves as Co-Editor-in-Chief of an international journal, and holds three patents. He has authored and edited six books on IoT, 5G, and AI, secured funding for two international projects, and delivered keynote presentations at over 40 international conferences. His certifications include Microservices Architecture, Data Science, and Foundations of AI, with research interests in Quality of Services in IoT-enabled devices, Neutrosophic Theory, and Neutrosophic Soft Set Systems.

Vijender Kumar Solanki is an Associate Professor in the Department of Artificial Intelligence & Data Science at the Stanley College of Engineering and Technology for Women, Hyderabad, TG, India, bringing over 10 years of academic experience in network security, IoT, Big Data, Smart City, and IT as a Senior Member of IEEE and an ACM member. Dr. Solanki has authored or co-authored more than 50 research articles published in various journals, books, and conference proceedings, and has edited or co-edited 14 books and conference proceedings in soft computing. He received his Ph.D. in Computer Science and Engineering from Anna University, Chennai, India, in 2017, an ME and MCA from Maharishi Dayanand University, Rohtak, Haryana, India, in 2007 and 2004, respectively, and a bachelor's degree in science from JLN Government College, Faridabad, Haryana, India, in 2001. Dr. Solanki serves as a Book Series Editor with CRC Press, Taylor & Francis Group, and is part of the Editorial Board for various international journals. and has been a guest editor with many other publishers.

Maurice E. Dawson is an Associate Professor of IT and Management and Director of the Center for Cyber Security and Forensics Education at Illinois Institute of Technology, Chicago, USA. He is a prominent figure in cybersecurity, IT, and data science with a career spanning several decades of significant contributions to academia, research, and industry. Dr. Dawson is widely recognized for his groundbreaking research in cybersecurity, data science, and digital forensics, addressing critical issues such as cyber threats, data breaches, and digital asset protection through numerous projects that bridge academia and industry. His collaborations with leading tech companies, government agencies, and research institutions have resulted in real-world impact solutions, earning him prestigious recognition, including the Fulbright Scholar Award. As an accomplished author, Dr. Dawson has published extensively in reputable academic journals, authored books, and contributed to numerous conference proceedings covering topics from cybersecurity risk management to data analytics and digital forensics.

Contributors

Sultan Ahmad
Department of Computer Science, College of Computer Engineering and Sciences
Prince Sattam Bin Abdulaziz University
Al-Kharj, Saudi Arabia
and
University Center for Research and Development (UCRD), Department of Computer Science and Engineering, Chandigarh University
Gharuan, Mohali, Punjab, India

Gufran Ahmad Ansari
University Center for Research and Development (UCRD), Department of Computer Science and Engineering
Chandigarh University
Gharuan, Mohali, Punjab, India

J. Chinna Babu
Department of Electronics and Communication Engineering
Annamacharya University
Rajampet, Andhra Pradesh, India

Sujit Bebortta
Department of Computer Science
Ravenshaw University
Cuttack, India

Ritu Raj Choudhary
Associate Professor, Faculty of Management and Commerce
Poornima University
Jaipur, Rajasthan, India

Khoa Nguyen Dang
Faculty of Engineering and Technology, International School
Vietnam National University
Hanoi, Vietnam

Maurice E. Dawson
Associate Professor, Information Technology and Management
Illinois Institute of Technology
Plainfield, Illinois

T. Kirthiga Devi
Research Scholar, Department of Computational Intelligence, School of Computing
SRM Institute of Science and Technology
Kattankulathur, India

Tran Quang Huy
Faculty of Physics
Hanoi Pedagogical University
Hanoi, Vietnam

Sudan Jha
Department of Computer Science & Engineering Kathmandu University Nepal
Dhulikhel, Kavre

Taarani Kaligotla
Department of Computer Science and Engineering
Stanley College of Engineering and Technology for Women
Hyderabad, India

Shaik Karimullah
Department of Electronics and Communication Engineering
Annamacharya University
Rajampet, Andhra Pradesh, India

K. Ravi Kishore
Department of Computer Science and Engineering
Stanley College of Engineering and Technology for Women
Hyderabad, India

Jackulin Mahariba A
Department of Computational Intelligence
SRM Institute of Science and Technology
Kattankulathur, Chennai, Tamil Nadu, India

Udit Mamodiya
Associate Professor & Associate Dean (Research), Faculty of Engineering and Technology
Poornima University
Jaipur, Rajasthan, India

G. Maragatham
Professor, Department of Computational Intelligence, School of Computing
SRM Institute of Science and Technology
Kattankulathur, India

Pham Tuan Minh
Faculty of Engineering and Technology, International School
Vietnam National University
Hanoi, Vietnam

Subhendu Kumar Pani
Krupajal Engineering College
Biju Patnaik University of Technology (BPUT)
Kausalya Ganga, Bhubaneswar
Odisha, India

Anwar Bhasha Pattan
Department of Electronics and Communication Engineering
BVRIT Hyderabad College of Engineering for Women
Hyderabad, Telangana, India

Deepak Prashar
School of Computer Science and Engineering
Lovely Professional University
Phagwara-Punjab, India

Satyananda Champati Rai
School of Computer Engineering
KIIT Deemed to be University
Bhubaneswar, Odisha, India

Manik Rakhra
School of Computer Science and Engineering
Lovely Professional University
Phagwara-Punjab, India

Salliah Shafi
GNA University
Phagwara, Punjab, India

Sarbagya Ratna Shakya
Department of Mathematical Sciences
Eastern New Mexico University
Portales, NM, USA

Vijender Kumar Solanki
Department of Artificial Intelligence & Data Science
Stanley College of Engineering and Technology for Women
Hyderabad, India

Prerna Srivastava
Associate Professor, Faculty of Management and Commerce
Poornima University
Jaipur, Rajasthan, India

Luong Thi Theu
Hoa Binh University
Hanoi, Vietnam

Duc-Tan Tran
Faculty of Electrical and Electronic Engineering
Phenikaa University
Hanoi, Vietnam

Biswajit Tripathy
Department of Computer Science & Engineering
BIT Mesra
Mesra, India

Niva Tripathy
Department of Computer Science
Driems University
Cuttack, India

Subhranshu Sekhar Tripathy
School of Computer Engineering
KIIT Deemed to be University
Bhubaneswar, Odisha, India

Tran Thi Van
Faculty of General Education
University of Labour and Social Affairs
Hanoi, Vietnam

Introduction

The start of the Fourth Industrial Revolution, or Industry 4.0, has ushered in an era of unprecedented connectivity and automation. The Industrial Internet of Things (IIoT) sits at the heart of this transformation, enabling cyber-physical systems to communicate, analyze data, and drive intelligent actions in real time. While this interconnectivity unlocks immense potential for efficiency and innovation, it simultaneously expands the attack surface, exposing critical industrial infrastructure to an increasingly sophisticated array of cyber threats. Traditional, perimeter-based security models are no longer sufficient to protect these complex, dynamic environments. This pressing challenge is the central focus of our book, *Cognitive Security for Industrial IoT: AI-Driven Approaches to Industrial 4.0.*

This volume is conceived at a critical juncture, where the need for adaptive, intelligent, and predictive security paradigms has never been more acute. The chapters within collectively argue that the solution lies in cognitive security—a paradigm that leverages the power of AI and ML to create defense mechanisms that are as dynamic and intelligent as the threats they aim to counter. The contributions explore a wide spectrum of AI-driven strategies, from foundational frameworks to cutting-edge applications, providing a holistic view of how cognitive systems can learn, predict, and respond to anomalies autonomously.

A key strength of this compilation is its balance between theoretical rigor and practical applicability. The authors, leading experts and researchers in their fields, delve into the architectural principles of secure IIoT systems and complement this with concrete case studies and implementation guidelines. Topics such as behavioral analytics, proactive threat hunting, and the security implications of emerging technologies like digital twins are examined in depth. This ensures the book is not just an academic exercise, but a vital resource for cybersecurity professionals, engineers, and policymakers tasked with safeguarding our industrial future.

Ultimately, this book serves as both a roadmap and a call to action. It outlines the current landscape of cognitive security, presents actionable solutions for today's problems, and thoughtfully speculates on the future challenges and ethical considerations of embedding AI into the very fabric of our industrial systems. We are confident that this collection will make a significant contribution to the field and become an essential reference for anyone invested in securing the promise of Industry 4.0.

Editors

1 Introduction to Cognitive Security in Industrial IoT

Sarbagya Ratna Shakya and Sudan Jha

1.1 INTRODUCTION

With advancements in the Internet of Things (IoT) technology in industrial sectors with Industry 4.0, the manufacturing industry has been revolutionized in recent years. According to Statista (www.statista.com/statistics/471264/iot-number-of-connected-devices-worldwide/), as shown in Figure 1.1, the number of IoT devices connected worldwide will be over 75 billion by 2025 [1]. IoT devices can connect with each other through the internet and are capable of collecting, processing, transmitting, and receiving data from each other [2]. IoT applied in Industry known as Industrial Internet of Things(IIoT)has been applied in many industrial sectors like manufacturing, oil and gas, life sciences[3], energy, transportation, agriculture, safety, and healthcare [4]. With valuable information of the manufacturing process, IIoT can efficiently allocate resources, optimize manufacturing process with automation, decrease cost, and integrate customer demand in the production with faster environmental adaptation [5]. Although utilizing the collected data from sensors, gateways, drivers, switches, cameras, robots, and control systems has increased operational efficiency, cost savings and improved safety, but added risks related to openness, scalability, interoperability and application independence [6]. Also with evolution of industry into Industry 4.0 that integrates knowledge for automation using various emerging technologies such as Machine learning (ML), artificial intelligence (AI), cyber-physical system (CPS), big data, fog computing, and edge computing [7].

Analyzing the insights from the collected data that provide intelligence has been proven to improve productivity and increase valuation of the company with increased revenue. With such big data collected from these IoT devices, concerns related to privacy, security, and trust have been encountered in this process. Hence with increasing applications, security and defending against cyberattacks have become critical and challenging and is becoming hard to address and maintain. Connectivity between IoT devices, servers, cloud, and third-party systems has increased the vulnerability to attackers. To protect the privacy and safety of the interaction of people with IoT technologies in industrial sectors requires systemic and cognitive security measures. With automation, the system and devices are more capable of taking decisions that adds reliability challenges.

DOI: 10.1201/9781003609773-1

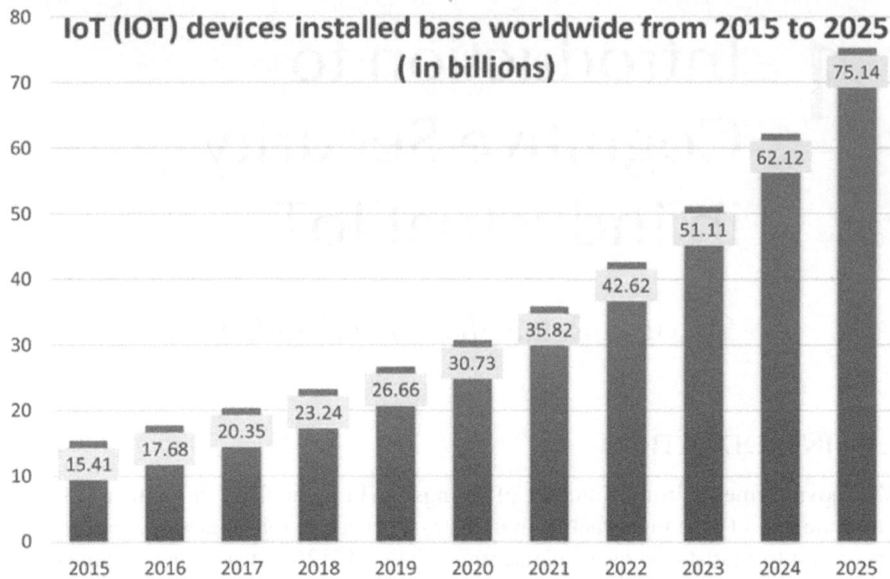

FIGURE 1.1 IoT connected devices installed base worldwide from 2015 to 2015.

With integration of AI and ML, with cybersecurity, cognitive security has been used in IIoT to create intelligent, self-learning systems that are capable of detecting, analyzing and responding to the cyber threats in real time [6]. They mimic human thinking using AI and ML to spot safety threats. It also uses AI and ML to automate threat detection and response. Analyzing vast amounts of data collected from different sensors and systems of the IIoT devices, cognitive security systems use advanced algorithms to identify patterns and anomalies to predict potential security threats in the systems. These systems analyze these vast amounts of past data, learn from them and adapt for new security threats. It can predict potential vulnerabilities and future threats. Cognitive security can also initiate appropriate responses like isolating the affected devices, rerouting network traffic, or updating security protocols, once a threat is detected. It protects again the online and offline malicious influence and helps in human decision making to solve problems.

1.2 IIOT COMMUNICATIONS

With various sensors, actuators, gateways, edge devices, and systems used in IIoT, the communication between these devices and systems are one of the critical sectors for effective application of IIoT. For data privacy, security and efficiency, edge computing can be applied for enhanced computational efficiency of the IoT and IIoT data security [8]. These components collect data, analyze, process, and generate insights from those data, and perform certain actions based on those insights. Also, these data need to be transmitted to the remote central systems or cloud for processing and storage. Hence, the exchange of data between these devices and systems without delay is important for seamlessly operating the process in real time. Also, protecting these data from unauthorized access and maintaining privacy are essential requirements to increase the adoption of IIoT and building trust.

Also, these data different communication protocols have been used like Industrial Ethernet, Wireless communications and low-power wide-area networks (LPWAN). Choosing the right protocol is extremely important and depends on the applications. The best protocol for IIoT applications depends on the factors such as range, power consumptions, data rate, reliability, and scalability.

Security of IIoT applications can be highlighted based on different IIoT layers. As shown in Figure 1.2, the IIoT layer architecture consists of three basic layers: application layer, network layer, and perception layer. The application layer consists of servers and clouds that consist of data. These layers are mainly vulnerable and could get ransomware attacks. With many IIoT devices connected to the internet and the companies cannot afford sudden downtime. These layers can be attacked and infected through malicious links and suspicious files. Apart from that several other attacks like malicious code injection attacks [10], cross-site attacks [11], malware injection attacks [12], data distortion attacks [13], SQL injection attacks and authorization and authentication attacks [14] are some of the major cyberattacks targeted in this layer. The second layer is the network layer, which provides the path for the message communication between the perception layer with the application layer. Various IIoT devices in the perception layer transmit data using a communication protocol through various communication links [15]. Some of the attacks like Sybil and ID cloning attacks [16], wormhole attacks [17], denial of service (DoS) attacks [18], and man in the middle attacks [19] are targeted toward this layer. The third layer is the perception layer that consists of end devices such as sensors, cameras, robots, and edge devices.

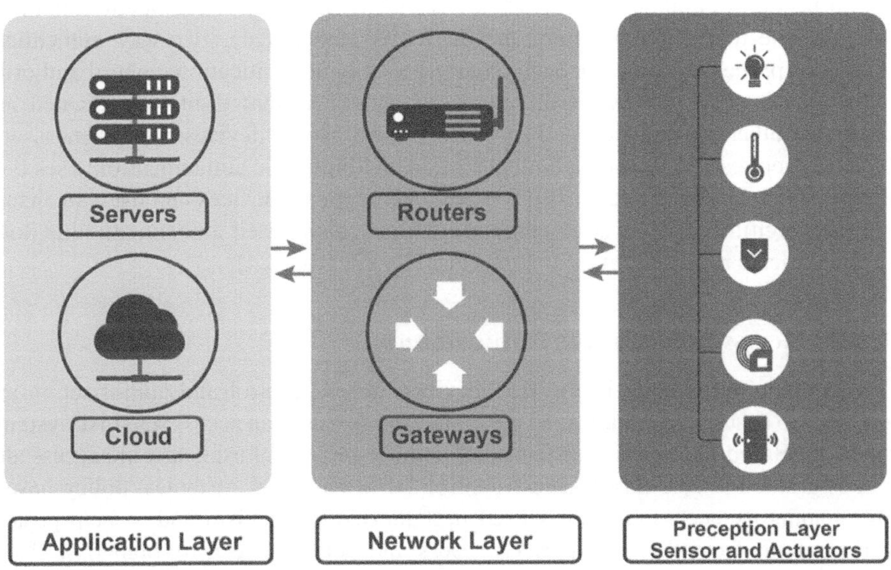

FIGURE 1.2 Three traditional IIoT layers [9].

The data collected through these devices are transmitted to the network layer using various communication mediums. Different attacks such as Node Capture attacks [20], jamming attacks, sleep deprivation attacks [21], replay attacks, [22] are targeted to the perception layers.

Different attacks can be launched at different layers for various purposes. The security of the data and networks is crucial among these different IIoT layers for continuous operations, efficiency, data privacy, and security. Different preventive measures can be applied in these various layers and security protocols can be applied for secure data transmission and applications.

1.3 IIOT SECURITY REQUIREMENTS

Along with the growth of IIoT integration in manufacturing industries and factories, the number of devices connected through the internet and its data collection has increased rapidly. Along with usage, IIoT security and privacy issues have also increased and have been on the major obligation for IIoT adoption. The security system required confidentiality, integrity, and availability as a building block for the security. Some of the security requirements for the implementation of IIoT are defined in this section.

1.3.1 Authentication

One of the key requirements for IIoT security is authentication that ensures only the authorized devices and system are accessed by only the authorized users. Also, the security and integrity of the data so that it doesn't alter during transit is another major concern. With limited power, limited storage and processing capacities, authentication becomes challenging in IIoT devices [23]. Different authentication methods such as one-way authentication: the device is authenticate, two-way authentication: both parties authenticate each other, three-way authentication: central authority authenticates both parties, certificate-based authentication; digital certificated are issued to authenticate devices, token-based authentication: devices uses tokens issued by an authentication server for authentication and biometric authentication: uses biometric data to authenticate users are used to authenticate devices and users to ensure both are legitimate. Also, edge computing, blockchain-based authentication is done to authenticate remote entities and provide data integrity [24].

1.3.2 Access Control and Authorization

For maintaining security in the IIoT devices, access control and authorization are essential to ensure only authorized personnel and devices can access sensitive systems and data. Based on the role such as administrator or normal user, specific access and role is granted in role-based access control. Also, based on attributes, policy-based, and mandatory access control is used in authorize mechanism and to help protect IIoT systems from the unauthorized access and potential security breaches. Using centralized managements, multifactor authentication, regular audits and updates, and network segmentation, access control and authorization are done to enhance the security of their IIoT systems.

1.3.3 Network and Device Security

Security of network and device is one of the key requirements in IIoT security. Both devices and network should be secured from cyber threats for ensuring the integrity and reliability of IIoT systems. Different methods such as segmentation of the network into smaller isolated segments or microsegments to limit the spread of potential attack are applied for network security. Using encryption, firewalls, implementing secure remote, and having regular network audits are used to identify and address vulnerabilities in the network. Also, for device security, implementing secure authentication, continuously monitoring device, and protecting from physical tampering is done for device security.

1.3.4 Resilience and Maintainability

Resilience and maintainability are one of the key requirements of IIoT security. The system should be resilience to continue its assigned tasks, and handle disruptions without significant loss of data or functionality. The system should be reliable, secure and trustworthy and should be able to self-heal and continue operating correctly even if some components of the system are infected. Also, maintainability refers as the capacity of the system to configure, reconfigure, and update a system or part of a system to be able to protect from previously known cyberattacks [25]. The update methods should be easier to update and replace the faculty components. Using predictive maintenance and training, issues should be predicted and managed and maintained in the system.

1.3.5 Privacy and Security

Privacy is another key requirement for IIoT security to protect highly sensitive data and information in IIoT. The vast collected data from the sensors and systems should be secured and protected ensuring confidentiality and integrity, hence maintaining trust. Different methods such as data minimization, encryption, differential privacy, and zero knowledge proofs are utilized to preserve the privacy of the sensitive data and ensuring compliance with the regulations. Also, system should collect data, maintaining transparency, its usage, and providing information about the data practice allowing useful information to be extracted from the data [26].

Along with data privacy, secure data sharing and security monitoring has also been an important prerequisite for IIoT. Monitoring security for early detection of security threats during data transfer and provide required response to these threats once identified in timely manner has been challenging [27]. Data availability has also been an important in IIoT for training and developing systems that can be used for smart maintenance, fault detection and prevention, optimization of the production, and cloud services. Hence, with these sensitive data, maintaining integrity, and availability, being able to securely use these intelligent and in efficient manner is required.

1.3.6 Real Time Threat Analysis

Real time threat analysis is another important key feature for cognitive security in IIoT. They involve using advanced technology that can monitor the network for possible threats, detecting those threats and once happen responding to those threats with necessary security measures. The technology uses ML and AI to analyze those vast amounts of data generated by IIoT devices identifying patterns and potential threats. The systems will monitor network traffic and device behavior to detect cyberattacks and use predictive maintenance, to perform preventive measures.

Hence, implementing security measures in IIoT based application should provide protection against cyber threats with real time threat analysis and prevention. Analyzing data collected through different IoT devices, learning patterns, and knowledge to predict potential cyber threats and implementing preventing measures to minimize the effects of the attacks is the primary requirements of the security measures in IIoT.

1.4 CATEGORIES OF IIOT SECURITY

Security threats have been one of the major challenges in IIoT applications. Modern cybersecurity is one of the effective ways to protect against them. The security of IIoT applications can be categorized based on three different components: device security, data security, and network security (Figure 1.3). Different attacks and their preventive measures can be explained based on these three security categories.

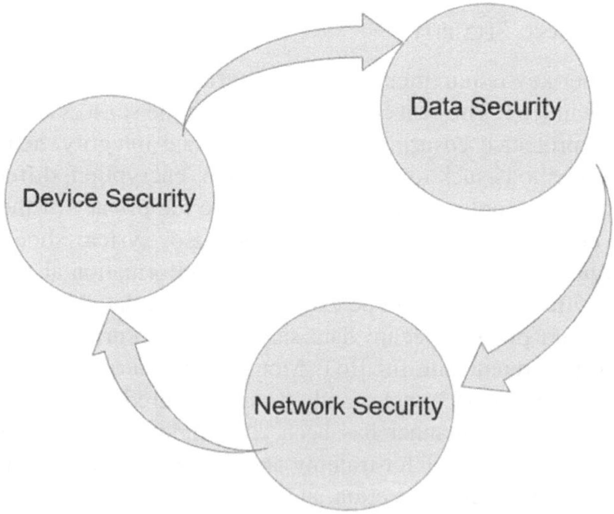

FIGURE 1.3 Categories of IIoT security.

1.4.1 Network Security

ML models have been used to monitor network traffic and identify unusual patterns to detect possible cybersecurity threats. It can be used to enhance network security by continuously analyzing data from the IIoT devices in real time. It is capable of detecting signal security breaches and network malfunctions. Also, using AI and ML models, they become capable of learning from past data and adapting to new threats. This helps to predict potential network or system failure and provide proactive approach to mitigate risks with preventive maintenance and hence providing minimum downtime [28]. Also, cognitive security provides and adapts security policies in real time and keeps evolving with the changing threat landscape.

1.4.2 Device Security

Cognitive security has a vital role in device security in IIoT. It can be implemented for secure boot processes that ensure only trusted devices and software are connected to the system. Also, different authentication and authorization mechanisms can be implemented to ensure that only those devices can interact with other IIoT devices and access the networks. By using network segmentation, in case of cyber threats, devices can be isolated from other devices of different segments of network.

1.4.3 Data Security

Cognitive security can be used to analyze real-time data collected from IIoT devices. Using AI and ML algorithms, these vast data can be analyzed, identify patterns, and anomalies for potential security threats. Different data-preserving technologies can be used for sensitive data protection during processing or the data transfer process. Algorithms like Grasshopper Black hole optimization [29] and zero trust architecture are used that enhance data privacy and security along with user and device verification protecting against cyber threats and reducing the risk of unauthorized access.

Hence to prevent unauthorized access, unclosed vulnerabilities, and malicious traffics in networks, different preventive measures like device authentication, data encryption, network segmentation, anomaly detection using ML, real-time threat intelligence, continuous monitoring, adaptive response mechanisms, secure lifecycle management, and collaboration with security operations centers can be implemented among these different categories of IIoT to prevent attack from hackers and cybercriminals.

1.5 DEVICE RELEVANT PROTOCOLS FOR COGNITIVE SECURITY IN IIOT

With high data transfer among different IIoT devices and networks, cognitive security protocols are required to ensure data security and integrity. To protect data and network connections, different protocols provide techniques and procedures that are required for data protection and prevent unauthorized or malicious attempts. Here are some lists of protocols applied in cognitive security in IIoT domains.

1.5.1 IIoT Safety and Security Protocol

IIoT [30] safety and security protocols provides detailed specifications and comprehensive guidelines for the design and implementation of the IIoT security systems. It also provides outlines for day-to-day operations, tools, and methods for accessing the security and performance and frameworks for solving security related issues and disputes.

1.5.2 MQTT (Message Queue Telemetry Transport)

MQTT [31] is one of the most efficient and reliable lightweight messaging protocol used in IIoT for remote monitoring and machine to machine connections. It uses a publish/subscribe model to allow for efficient communication between devices. The publishers send messages to topics and the subscribers receive that message from a central broker. This makes it possible to communicate between two devices without direct device to device interactions. This protocol is designed for resource constraints devices and networks allowing for real time monitoring and control. It also supports encryption and authentication mechanisms for data security and access control.

1.5.3 CoAP (Constrained Application Protocol)

CoAP protocol [32] is designed for constrained networks and devices in IIoT. It uses RESTFUL architecture, like Hyper Text Transfer Protocol (HTTP), and is used for web transfer protocol. It is simple and lightweight with a small message overhead and is suitable for low power, lossy and resource constraint networks. CoAP message consists of header that consists of message type, message ID, token, and options information and payload that carries the actual data being transmitted. It provides low overhead, a simple request-response model, and is suitable for applications with low network bandwidth.

1.5.4 DTLS (Datagram Transport Layer Security)

DTLS is designed to provide security for datagram-based applications in IIoT. This protocol was derived from Transport Layer Security (TLS) and provides encryption, integrity and authenticity. It provides low overhead and is suitable for lossy networks. It provides end-to-end security, hence preventing tampering, message forgery and eavesdropping.

1.5.5 XMPP (Extensible Messaging and Presence Security)

XMPP [33, 34] is designed for fast, near real time exchange of data between multiple entities on a network. It supports publish/subscribe models like MQTT and facilitates efficient data distribution. It provides robust security features, with encryption with TLS and authentication mechanisms like SASL for secure communication and access control. XMPP extension like XEP -0324 provides efficient provisioning of services and access rights in IIoT environments [35].

Different applications have been applied in various use cases such as industrial automation, smart manufacturing, device management, environmental monitoring, and real time monitoring and control. These protocols provide frameworks for secure and efficient communication in different layers enhancing communication. Security, interoperability, and scalability in IIoT.

1.6 BENEFITS OF COGNITIVE SECURITY IN IIOT

With advanced technology like AI and ML, implementing cognitive security will enhance the security system of IIoT systems. With ability to learn from past data and updating itself to be able to detect mitigating vulnerabilities in real time will help to protect IoT devices from evolving cyber threats. Some of the major benefits of applying cognitive security for IIoT security are described in the next section.

1.6.1 Improved Threat Detection and Response Time

Cognitive security enhances the threat detection and response time in IIoT through several mechanisms. With advanced ML and vast data from IIoT devices, cognitive security systems can identify patterns and anomalies that help them to detect potential security threats. These systems learn from the ongoing traffic and keep on improving threat detection capabilities through adaptive learning. Once the threat is detected, cognitive security systems can perform real time analysis of network traffic and initiate appropriate response to the threat automatically. This automation reduces response time for the decision making and provides the latest threats information analyzing new patterns that could indicate emerging risks. Also, integrating security with edge computing helps reduce latency and allows faster detection and response with quicker insights and actions.

1.6.2 Improved Ability to Handle Complex, Evolving Cyber Threats

With pattern learning and adaptive learning capabilities, analyzing historical data, cognitive security systems can predict potential future threats. By analyzing real time patterns, anomalies, and identifying unusual behavior, they can detect potential threats early or even before they occur. This helps to improve response time to complex threats and can also provide immediate alerts and act quickly to mitigate risks. With the latest data points and insights, they assist humans with recommendations to effectively detect and response to the most recent cyberattacks. With adaptive threat detection and prevention, the system keeps learning and evolving with the latest data points and insights and improves over time.

1.6.3 Reduced Human Error in Security Operations

With automated threat detection systems, AI powered cognitive security systems can repeatedly and routinely perform security tasks operation minimizing human errors. They can analyze vast amounts of data, monitor network traffic, log analysis, vulnerability scanning, patch management, update security, manage access control, and identify potential threats more accurately and quickly than humans. With real time analysis, detection and response to threats, it reduced human errors in critical moments. Also, by reducing high cognitive load by handling large complex data, and complex analysis, it helps human operators and allow them to focus on more strategic tasks, and reducing fatigue related errors [36].

1.6.4 Scalability to Protect Large Networks of IIoT Devices

Cognitive security systems can automatically initiate appropriate responses to mitigate risks across many devices simultaneously and efficiently. This reduces the need for manual intervention and provides security operations to handle large number of devices and incidents even when the network expands, improving scalability in threat management. Also, since cognitive security employs adaptive algorithms, these algorithms dynamically allocate resources and be able to adjust to the changing scale and complexity of the network. With ML models, they can continuously relearn and improve to detect new and evolving threats [37]. Cognitive security platforms provides a uniform, distributed security mechanisms that can efficiently detect attacks at a faster rate in highly scalable IoT environments and enhances scalability by efficiently handling real time processing demands [38].

1.7 CHALLENGES

Cognitive security in Industrial IoT faces several challenges such as scalability, integration with existing IoT devices, and lack of standardization and skilled workforce. With increasing old seniors, IIoT devices, and systems that are not designed to function safely online, it forms a huge potential attack surface. Also, implementing cognitive security in IIoT can be very costly with unclear return on investment (ROI). Lack of cybersecurity technology and practices, outdated firmware, and non-updatable software have added more threats in cyberattacks and adoption of cognitive security in IIoT [39].

1.7.1 Infrastructure

Many industrial system have traditional equipment and protocols that may not support actions for modern cybersecurity threats. These IIoT networks can consist of a mixture of modern devices along with these legacy systems, which also adds challenges creating complex environments and networks where each device can serve as an entry point for potential attacks. Also, integrating AI with the existing industrial control systems can be complex, costlier and requires significant technical expertise and skilled manpower.

1.7.2 Scalability Issues

Scalability has been one of the major issues in applying cognitive security in IIoT. With massive volume of data generated from various sensors and IoT devices, managing, processing and securing these vast data has been a major challenge. With new devices and network connection increasing day by day, maintaining threat detection and managing security across vast diverse IIoT devices has been difficult. With different heterogeneous devices and systems, cognitive security solutions should address interoperability issues and ensure seamless integration across the network [40]. Also, with limited computational power and memory, and resource constraint, developments of robust cognitive security and its implementation in many IIoT devices has been difficult and limited.

1.7.3 Lack of Standardized Security Protocols

Due to the lack of comprehensive and standardized security protocols for cognitive security in IIoT, and security gaps between different devices and systems that work on different levels of security, it adds challenges to ensure comprehensive security across the entire IIoT network. Although there are some standards but is not sufficient to implement consistent security measures which lead to interoperability issues and vulnerabilities to cyber threats. Hence widespread adoption and compliance has been challenging with potentially much more complicated legal and reputational risks. This has also added complications in implementing and maintenance of the cognitive security systems.

1.7.4 Quality Data

Since the effectiveness of the AI models depends highly on the data used for training, it adds challenges in the availability of quality industrial data for cognitive security. Although various sources and sensors of IIoT systems can provide large amounts of data, ensuring the quality of these heterogeneous data will be challenging [9].

1.7.5 Privacy Concerns

With the use of legacy and insecure communication systems and lack of data protection measures, maintaining data privacy has been one of the major challenges for cognitive security in IIoT devices. With the vast sensitive data collected from various IIoT devices, ensuring its security along with allowing extraction of useful insights from that data is complex. With regulatory compliance and operational needs, balancing privacy with the operational requirements of IIoT systems adds significant challenges in IIoT domains [41].

Although implementing cognitive security measures in IIoT applications still faces many challenges, with a multifaceted approach, these challenges can be overcome. Applying these security measures will help industries to protect their IIoT systems and ensure the safety of their operations.

1.8 CONCLUSION

IIoT connects different devices and networks and uses the insights of these data, which helps industries in their decision making and optimizing manufacturing and improve productivity. Cognitive security can be adopted for security, privacy, and availability of these large sensitive data, data transfer and network. In this chapter, we discussed the basics of communication and discussed different cyberattacks into three traditional IIoT layers: application layer, network layer, and perception layer. Edge devices that are continuously collecting data and transferring it to application layers through the network layers. The security threats that can cause system downtime, compromise in data and network security and privacy should be neutralized and prevented. Also, we discussed the security requirements based on network security, device security, and data security. Different protocols that provide the guidelines and architecture for the security implementation in the IIoT are discussed. Additionally, the key benefits of applying cognitive security such as real-time threat detection, scalability, minimizing human errors and ability to upgrade with new cyber threats in IIoT domain was discussed. Implementing security upgrades in industrial sectors still encountered various challenges. Lack of advanced infrastructure, high cost for scalability, lack of standard protocols that can be applied for all, lack of quality data to developed new system and privacy issues are some of the major challenges for adopting cognitive security in IIoT.

REFERENCES

[1] I. Adascalitei, "Smartphones and IoT Security," *Inform. Econ.*, vol. 23, pp. 63–75, 2019, doi: 10.12948/issn14531305/23.2.2019.06.

[2] M. B. Shishehgarkhaneh, R. C. Moehler, and S. F. Moradinia, "Blockchain in the Construction Industry Between 2016 and 2022: A Review, Bibliometric, and Network Analysis," *Smart Cities*, vol. 6, no. 2, pp. 819–845, 2023.

[3] S. R. Shakya and S. Jha, "Chapter 9 – Internet of Things Application in Life Sciences," in *Internet of Things in Biomedical Engineering*, V. E. Balas, L. H. Son, S. Jha, M. Khari, and R. Kumar, Eds., Academic Press, 2019, pp. 213–233. doi: 10.1016/B978-0-12-817356-5.00012-7.

[4] H. Jaidka, N. Sharma, and R. Singh, "Evolution of IoT to IIoT: Applications & Challenges," in Proceedings of the international conference on innovative computing & communications (ICICC), 2020.

[5] H. Xu, W. Yu, D. Griffith, and N. Golmie, "A Survey on Industrial Internet of Things: A Cyber-Physical Systems Perspective," *IEEE Access Pract. Innov. Open Solut.*, vol. 6, 2018. doi: 10.1109/access.2018.2884906.

[6] S. Jha, L. Nkenyereye, G. P. Joshi, and E. Yang, [IoT6], "Mitigating and Monitoring Smart City Using Internet of Things," *CMC-Computers, Materials & Continua*, vol. 65, no. 2, pp. 1059–1079, 2020. doi: 10.32604/cmc.2020.011754.

[7] D. Stefanescu, P. Galán-García, L. Montalvillo, J. Unzilla, and A. Urbieta, "Industrial Data Homogenization and Monitoring Scheme with Blockchain Oracles," *Smart Cities*, vol. 6, pp. 263–290, 2023.

[8] J. Ni, X. Lin, and X. S. Shen, "Toward Edge-Assisted Internet of Things: From Security and Efficiency Perspectives," *IEEE Netw.*, vol. 33, no. 2, pp. 50–57, 2019.

[9] B. Alotaibi, "A Survey on Industrial Internet of Things Security: Requirements, Attacks, AI-Based Solutions, and Edge Computing Opportunities," *Sensors*, vol. 23, no. 17, 2023, doi: 10.3390/s23177470.

[10] A. S. Choudhary and M. L. Dhore, "CIDT: Detection of Malicious Code Injection Attacks on Web Application," *Int. J. Comput. Appl.*, vol. 52, no. 2, pp. 19–26, 2012.

[11] N. Gelernter and A. Herzberg, "Cross-Site Search Attacks," in Proceedings of the 22nd ACM SIGSAC Conference on Computer and Communications Security, 2015, pp. 1394–1405.

[12] R. Madhvan and M. F. Zolkipli, "An Overview of Malware Injection Attacks: Techniques, Impacts, and Countermeasures," *Borneo Int. J.*, vol. 6, no. 3, pp. 22–30, 2023.

[13] M. M. N. Aboelwafa, K. G. Seddik, M. H. Eldefrawy, Y. Gadallah, and M. Gidlund, "A Machine-Learning-Based Technique for False Data Injection Attacks Detection in Industrial IoT," *IEEE Internet Things J.*, vol. 7, no. 9, pp. 8462–8471, 2020.

[14] A. C. Panchal, V. M. Khadse, and P. N. Mahalle, "Security Issues in IIoT: A Comprehensive Survey of Attacks on IIoT and its Countermeasures," in 2018 IEEE Global Conference on Wireless Computing and Networking (GCWCN), 2018, pp. 124–130.

[15] L. Wallgren, S. Raza, and T. Voigt, "Routing Attacks and Countermeasures in the RPL-Based Internet of Things," *Int. J. Distrib. Sens. Networks*, vol. 9, no. 8, p. 794326, 2013.

[16] S. Jha, G. P. Joshi, L. Nkenyereya, D. W. Kim, and F. Smarandache, [IoT8], "A Direct Data-Cluster Analysis Method Based on Neutrosophic Set Implication," *CMC-Computers, Materials & Continua*, vol. 65, no. 2, pp. 1203–1220, 2020. doi: 10.32604/cmc.2020.011618.

[17] Y.-C. Hu, A. Perrig, and D. B. Johnson, "Wormhole Attacks in Wireless Networks," *IEEE J. Sel. Areas Commun.*, vol. 24, no. 2, pp. 370–380, 2006.

[18] Q. Gu and P. Liu, "Denial of service attacks," *Handb. Comput. Networks Distrib. Networks, Netw. Planning, Control. Manag. New Trends Appl.*, vol. 3, pp. 454–468, 2007.

[19] H. A. Abdul-Ghani, D. Konstantas, and M. Mahyoub, "A Comprehensive IoT Attacks Survey Based on a Building-Blocked Reference Model," *Int. J. Adv. Comput. Sci. Appl.*, vol. 9, no. 3, pp. 355–373, 2018.

[20] M. Niedermaier, F. Fischer, D. Merli, and G. Sigl, "Network Scanning and Mapping for IIoT Edge Node Device Security," in 2019 International Conference on Applied Electronics (AE), 2019, pp. 1–6.

[21] M. Pirretti, S. Zhu, N. Vijaykrishnan, P. McDaniel, M. Kandemir, and R. Brooks, "The Sleep Deprivation Attack in Sensor Networks: Analysis and Methods of Defense," *Int. J. Distrib. Sens. Networks*, vol. 2, no. 3, pp. 267–287, 2006.

[22] J. Cecílio and A. Souto, "Security Issues in Industrial Internet-of-Things: Threats, Attacks and Solutions," in 2024 IEEE International Workshop on Metrology for Industry 4.0 & IoT (MetroInd4. 0 & IoT), 2024, pp. 458–463.

[23] T. Pereira, L. Barreto, and A. Amaral, "Network and Information Security Challenges within Industry 4.0 Paradigm," *Procedia Manuf.*, vol. 13, pp. 1253–1260, 2017, doi: 10.1016/j.promfg.2017.09.047.

[24] F. Wang, J. Cui, Q. Zhang, D. He, C. Gu, and H. Zhong, "Blockchain-Based Lightweight Message Authentication for Edge-Assisted Cross-Domain Industrial Internet of Things," *IEEE Trans. Dependable Secur. Comput.*, vol. 21, no. 4, pp. 1587–1604, 2024. doi: 10.1109/TDSC.2023.3285800.

[25] L. Zhou and H. Guo, "Anomaly Detection Methods for IIoT Networks," in *2018 IEEE International Conference on Service Operations and Logistics, and Informatics (SOLI)*, IEEE Press, 2018, pp. 214–219. doi: 10.1109/SOLI.2018.8476769.

[26] V. Demertzi, S. Demertzis, and K. Demertzis, "An Overview of Privacy Dimensions on the Industrial Internet of Things (IIoT)," *Algorithms*, vol. 16, no. 8, 2023, doi: 10.3390/a16080378.

[27] M. F. Elrawy, A. I. Awad, and H. F. A. Hamed, "Intrusion Detection Systems for IoT-Based Smart Environments: A Survey," *J. Cloud Comput.*, vol. 7, no. 1, pp. 1–20, 2018.

[28] G. V. Barekar, R. Purandare, A. Sawlikar, R. R. Welekar, P. K. Ingole, and N. Shelke, "Enhancing Security and Reliability in Industrial IoT Networks through Machine Learning," *J. Electr. Syst.*, vol. 20, no. 1s, pp. 289–302, 2024.

[29] M. Kumar, R. L. Shankar, V. P. Mishra, R. K. Dhanaraj and B. Balusamy, "A Smart Privacy Preserving Framework for Industrial IoT Using Hybrid Meta-Heuristic Algorithm," *Sci. Rep.*, vol. 13, no. 1, p. 5372, 2023.

[30] S. Figueroa-Lorenzo, J. Añorga, and S. Arrizabalaga, "A Survey of IIoT Protocols: A Measure of Vulnerability Risk Analysis Based on CVSS," *ACM Comput. Surv.*, vol. 53, no. 2, pp. 1–53, 2020.

[31] B. Arbab-Zavar, E. J. Palacios-Garcia, J. C. Vasquez, and J. M. Guerrero, "Message Queuing Telemetry Transport Communication Infrastructure for Grid-Connected AC Microgrids Management," *Energies*, vol. 14, no. 18, 2021. doi: 10.3390/en14185610.

[32] Z. Shelby, K. Hartke, and C. Bormann, "The Constrained Application Protocol (CoAP)," no. 7252. in *Request for Comments*. RFC Editor, Jun. 2014. doi: 10.17487/RFC7252.

[33] P. Saint-Andre, "Extensible Messaging and Presence Protocol (XMPP): Core," no. 6120. in *Request for Comments*. RFC Editor, Mar. 2011. doi: 10.17487/RFC6120.

[34] A. A. Alli and M. M. Alam, "The Fog Cloud of Things: A Survey on Concepts, Architecture, Standards, Tools, and Applications," *Internet of Things*, vol. 9, p. 100177, 2020. doi: 10.1016/j.iot.2020.100177.

[35] S. Jha, C. Seo, E. Yang, and G. P. Joshi, "Real-time object detection and tracking system for video surveillance system," *Multimedia Tools and Applications*, vol. 80, no. 3, pp. 3981–3996, 2021. doi:10.1007/s11042-020-09749-x.

[36] F. M. Cau and L. D. Spano, "Mitigating Human Errors and Cognitive Bias for Human-AI Synergy in Cybersecurity," in *DAMOCLES@AVI*, 2024. [Online]. Available: https://api.semanticscholar.org/CorpusID:270880020

[37] P. R. Agbedanu, S. J. Yang, R. Musabe, I. Gatare, and J. Rwigema, "A Scalable Approach to Internet of Things and Industrial Internet of Things Security: Evaluating Adaptive Self-Adjusting Memory K-Nearest Neighbor for Zero-Day Attack Detection," *Sensors*, vol. 25, no. 1, 2025. doi: 10.3390/s25010216.

[38] S. Prabavathy and V. Supriya, "SDN Based Cognitive Security System for Large-Scale Internet of Things using Fog Computing," in 2021 International Conference on Emerging Techniques in Computational Intelligence (ICETCI), 2021, pp. 129–134. doi: 10.1109/ICETCI51973.2021.9574051.

[39] S. R. Shakya and S. Jha, "Challenges in Industrial Internet of Things (IIoT)," in *Industrial Internet of Things*, CRC Press, 2022, pp. 19–39.

[40] E. Bajramovic, D. Gupta, Y. Guo, K. Waedt, and A. Bajramovic, "Security Challenges and Best Practices for IIoT," in *INFORMATIK 2019: 50 Jahre Gesellschaft für Informatik – Informatik für Gesellschaft (Workshop-Beiträge)*, Bonn, 2019, pp. 243–254, ISBN: 978-3-88579-689-3. doi: 10.18420/inf2019_ws28.

[41] C.-C. Lin, C.-T. Tsai, Y.-L. Liu, T.-T. Chang, and Y.-S. Chang, "Security and Privacy in 5g-IIoT Smart Factories: Novel Approaches, Trends, and Challenges," *Mob. Networks Appl.*, vol. 28, no. 3, pp. 1043–1058, 2023.

2 A Comprehensive Review on Challenges, Frameworks, and Future Trends for Intelligent Security in Industrial IoT

Shaik Karimullah, J. Chinna Babu, Anwar Bhasha Pattan, Sultan Ahmad, and Sudan Jha

2.1 INTRODUCTION

The rapid expansion of Internet of Things (IoT) devices across industrial and smart city applications has revolutionized the way we interact with technology, providing a host of benefits that include increased automation, enhanced data collection, and greater operational efficiency [1]. IoT has enabled real-time monitoring of systems, better resource management, and the ability to make informed decisions based on vast amounts of collected data. In industrial settings, for example, IoT devices can optimize manufacturing processes, reduce downtime, and enhance safety. Similarly, in smart cities, IoT plays a pivotal role in improving traffic management, waste management, energy consumption, and even public safety. These innovations are transforming the fabric of modern society, making processes smarter, more responsive, and cost-effective [2].

However, as IoT systems become more ubiquitous and interconnected, they also introduce significant security challenges. The more devices that are connected to a network, the more potential vulnerabilities there are for attackers to exploit. Each IoT device, from sensors to smart appliances, can become a potential entry point for cyberattacks, leading to an increased risk of data breaches, unauthorized access, and the compromise of personal or organizational information. For example, a breach in an industrial IoT system could result in the disruption of critical processes or expose sensitive manufacturing data [3]. In smart cities, a security breach in IoT systems can jeopardize public safety, with hackers potentially manipulating traffic lights, surveillance systems, or public utilities. As a result, the security of IoT systems has become one of the most critical concerns for both users and organizations [4]. The potential consequences of failing to secure IoT devices—ranging from financial loss to reputational damage—make it essential to develop and implement robust security measures that ensure the integrity, privacy, and functionality of IoT systems. Recent

DOI: 10.1201/9781003609773-2

research efforts have made significant strides in addressing these security concerns by incorporating a variety of advanced techniques, including machine learning (ML), encryption, and blockchain technologies. ML, for example, has emerged as a powerful tool for detecting anomalies in IoT networks. Bharath Reedy Konatham et al. [5] introduced a hybrid ML model combining Convolutional Neural Networks (CNNs) and Gated Recurrent Units (GRUs) for anomaly detection, showing considerable improvements in detection accuracy [6]. Their model not only increased the precision of identifying real threats but also minimized false alarms, which is particularly important in industrial applications where unnecessary interruptions could lead to costly delays [7]. This approach leverages ML's ability to process large datasets and identify patterns of normal and abnormal behavior, making it an essential tool for real-time security monitoring in IoT environments as shown in Figure 2.1.

In addition toML, encryption plays a vital role in securing IoT systems. Liming Fang et al. (2020) proposed a novel security scheme for industrial IoT systems, which used conditional proxy re-encryption to safeguard data. This approach allows data to be securely shared across different IoT platforms without exposing sensitive information to unauthorized entities [8]. By focusing on fine-grained data access control and low computational overhead, this scheme makes it easier to deploy security measures without compromising the performance of IoT systems. Such encryption methods ensure that data remains confidential and that only authorized parties can access specific information, thus enhancing both security and privacy within IoT networks. Meanwhile, blockchain technology has emerged as a robust solution for securing IoT environments, particularly in terms of authentication and data integrity. A 2022 study explored a blockchain-based authentication mechanism that combined elliptic curve cryptography (ECC) and public key infrastructure (PKI) to ensure secure communication between IoT devices. Blockchain's decentralized nature helps mitigate common IoT security threats such as data leakage, eavesdropping, and

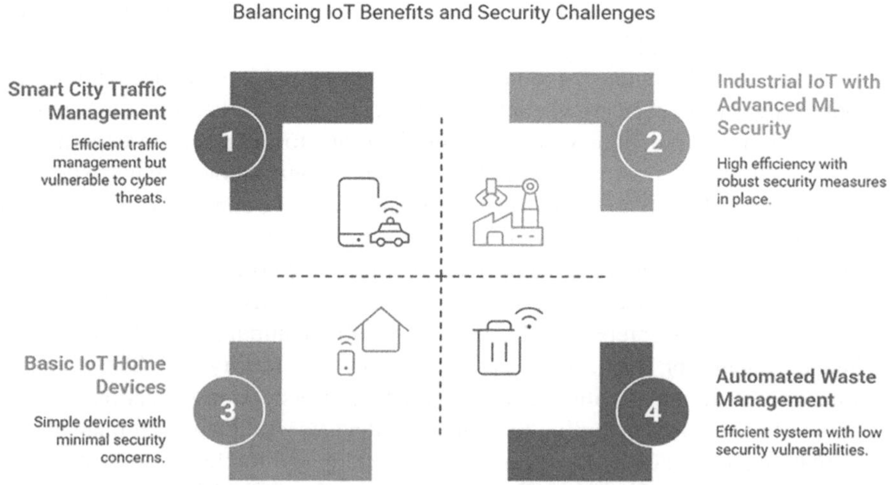

FIGURE 2.1 Balancing benefits and security challanges.

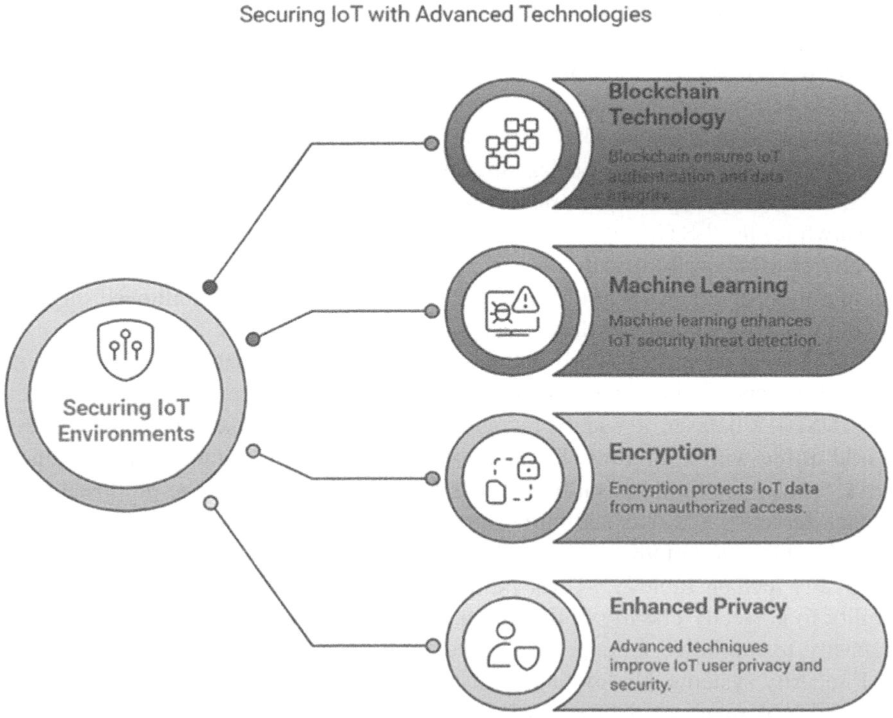

FIGURE 2.2 Securing IoT with advanced technologies.

replay attacks, by providing a transparent and immutable ledger of transactions. This ensures that every interaction between devices is verified and recorded in a tamper-resistant manner, reducing the risk of cyberattacks and enhancing privacy. As IoT systems become more complex and interconnected, blockchain's ability to offer a secure and decentralized way to manage authentication and transactions makes it a critical technology for safeguarding IoT environments. Together, these studies highlight the transformative potential of combining ML, encryption, and blockchain technologies to address the complex security challenges of IoT systems. By leveraging these advanced techniques, researchers are developing more resilient, secure, and scalable solutions to protect IoT devices and networks from emerging threats. As the IoT landscape continues to evolve, these technologies offer a promising path forward in ensuring the integrity, privacy, and functionality of IoT ecosystems, providing greater confidence for both users and organizations that rely on these systems for their everyday operations as shown in Figure 2.2.

2.2 COMPARATIVE ANALYSIS OF EXISTING FRAMEWORKS

N. Patel et al., in 2022, published the article titled "Deep Learning based Intrusion Detection in Internet of Things using Convolutional Neural Network and Long Short-Term Memory." This research addressed the pressing issue of security in IoT environments, which are increasingly vulnerable to sophisticated cyber-attacks due to their heterogeneous and distributed nature. The authors proposed an innovative hybrid deep learning model that integrates CNNs with Long Short-Term Memory (LSTM) networks to enhance the accuracy and responsiveness of intrusion detection systems (IDS) in IoT networks. As per the table 2.1, the CNN component was utilized for its excellent capability in extracting spatial features from the input network traffic data, effectively capturing intricate patterns that may signal malicious behavior. Meanwhile, the LSTM component, a type of recurrent neural network known for its ability to retain information across long sequences, was employed to understand temporal dependencies within the traffic flow. This combination allowed the model to analyze not just isolated data packets, but also the sequence and behavior of data over time, making it particularly adept at detecting sophisticated and evolving threats.

To train and evaluate the performance of their model, the authors employed the CICIDS2017 dataset, a widely recognized benchmark dataset that simulates real-world traffic with a mixture of normal behavior and various attack types, including DoS, brute-force, infiltration, botnet activity, and web attacks. The proposed hybrid model demonstrated outstanding performance, achieving an impressive accuracy of 99.2%, a precision of 98.6%, a recall of 97.8%, and an F1-score of 98.2% which can be clearly seen on Table 2.2. These performance metrics not only indicate the model's ability to correctly classify both benign and malicious traffic, but also highlight its capacity to minimize false positives and false negatives—key factors in real-time IoT security systems. Nikhil Gaurav et al., in 2022, presented a comprehensive and forward-looking research article titled "Security in Cognitive IoT: Opportunities and Challenges." This scholarly work delved into the emerging paradigm of Cognitive Internet of Things (CIoT), which represents the convergence of cognitive computing capabilities—such as artificial intelligence (AI), ML, and context-aware decision-making—with the vast, interconnected ecosystem of IoT devices. The authors aimed to address the increasingly complex and dynamic security challenges faced by traditional IoT frameworks, which often struggle to cope with the scale, heterogeneity, and sophistication of modern cyber threats.

In their detailed exposition, the authors first introduced the concept of CIoT and highlighted its transformative potential in enabling intelligent, autonomous, and self-adaptive systems. They emphasized that while traditional IoT architectures primarily rely on static rule-based models and cloud-dependent processing, CIoT systems can dynamically learn from their environments and make proactive decisions through distributed intelligence. However, this evolution also introduces novel security vulnerabilities, as the cognitive layer itself becomes a potential attack vector. This chapter identified and critically analyzed key security issues plaguing CIoT systems—chief among them being data privacy, secure data transmission, identity management, trust establishment, and resilience to adversarial ML attacks. The authors argued that due to the sensitive and continuous data flows involved in CIoT applications such as smart healthcare, intelligent surveillance, and personalized smart homes, ensuring end-to-end data protection is paramount. They also explored how the context-awareness of cognitive agents can be exploited to develop more nuanced security mechanisms capable of real-time adaptation to evolving threats.

TABLE 2.1
Significant Contributions

S.No	Authors	Year	Methodology	Insights	Results	Conclusions
1	Konstantinos Loupos et al.	2019	Cognitive security framework using IPSE, blockchain, ML-based anomaly detection	Emphasizes real-time safety/security in IIoT via risk mapping and closed-loop ML monitoring	Integrated blockchain for secure data, real-time risk mapping using IoT topology	Demonstrated an intelligent supervision platform to enhance safety and security in industrial IoT environments
2	Wenying Zheng et al.	2021	Secure storage auditing with efficient key updates, hybrid data dynamics in cognitive IIoT	Addresses data integrity and key management in semi-trust cloud environments	Achieved 40% faster data block location, verified under diffie-Hellman and DLog assumptions	Enhanced storage auditing efficiency and security for cognitive industrial IoT environments
3	Jiani Fan et al.	2023	Differentiated security architecture with incentive mechanisms for cognitive IoT	Tackles cognitive IoT evolution and secure P2P communication	Proposed secure and efficient data communication system	Future work focuses on secure incentive-based data sharing using deep learning
4	Hsing-Chung Chen et al.	2021	Online cognitive authentication and trust evaluation API using modified EPC and CSG	Ensures secure communication among unknown IoT devices in distributed networks	Used EPC tags for identity and three cognitive sub-APIs	Suitable for enabling mesh trust networks in industrial cognitive IoT systems
5	Fardin Jalil Piran et al.	2023	Hyperdimensional cognitive computing for IIoT cyberattack detection	Introduces brain-like computing for intrusion detection with limited data and training cycles	Achieved 92.6% accuracy with only 300 samples and 5 iterations	Outperformed traditional ML in IIoT security, showing promise for low-sample environments

(*continued*)

TABLE 2.1 (Continued)
Significant Contributions

S.No	Authors	Year	Methodology	Insights	Results	Conclusions
6	Hanlin Chen et al.	2019	Security protection framework focusing on acquisition, transport, and processing layers	Identifies vulnerabilities in MQTT protocol and botnet virus risks	Designed multi-layer security strategy	Emphasized tailored solutions at each layer of industrial IoT systems
7	Kushal Kumar B.N, M. Panduranga Rao	2017	Cognitive intrusion detection system using data acquisition and central monitoring	Proposes lightweight, adaptive detection suited for resource-constrained industrial environments	Central system diagnoses attacks effectively	Anticipates better performance than traditional security frameworks when implemented
8	Gabriel Portal et al.	2020	Edge security architecture using blockchain and context-aware mechanisms	Argues against adapting cloud/embedded methods for IoT; calls for dedicated architectures	Validated in Additive Manufacturing Units use case	Blockchain + context-aware architecture is viable for industrial IoT threat mitigation
9	Emrah Tomur et al.	2021	Taxonomy of threats; scenario-based analysis	Focus on interplay between security vulnerabilities and functional safety in Industrial IoT	Identified attack scenarios on physical and cyber-assets	Recommended mitigations based on standards and mobile tech features
10	Ulla Gain	2021	Framework integration and business model evaluation	Emphasized cognitive services for enhanced transparency and insight	Data-driven strategies improve IIoT value propositions	Cognitive services support security via intelligent collaboration
11	Farzeen Basith et al.	2024	Review of CIoT with AI/ML; focus on security & ethics	Explores secure, ethical AI in CIoT systems	Highlighted privacy, fairness, transparency needs	Responsible AI critical for sustainable CIoT systems

12	Arbia Riahi et al.	2014	Systemic cognitive model (triangular pyramid)	Interaction analysis among people, processes, objects	Security policies enhance safety and reliability	Cognitive approach vital for holistic IoT security
13	Liming Fang et al.	2020	Conditional proxy re-encryption scheme	Universal model for IIoT scenarios; secure access control	Achieved functionality & security with low overhead	Scheme ensures secure data in IIoT platforms
14	Bharath Reedy Konatham et al.	2023	Deep learning (CNN, GRU, LSTM, hybrid CNN+GRU)	Comparative anomaly detection study	Hybrid model: 94.94% accuracy, 0.001 FAR	Hybrid model effective, trade-off with convergence time
15	Anju Devi, Geetanjali Rathee & Hemarj Saini	2022	Hybrid blockchain authentication scheme	Combines ECC & PKI using private blockchain	Prevented data leakage, DoS, eavesdropping, etc.	Verified secure authentication in IIoT systems
16	Lehlogonolo P. I. Ledwaba, Gerhard P. Hancke	2021	IIoT layer-based security strategy	Exposure of critical systems to cyber threats	Identified key services per IIoT stack layer	Comprehensive security essential for Industry 4.0
17	Andrea Sánchez-Zumba, Diego Ávila-Pesantez	2023	STRIDE model analysis on 43 studies	Review of IIoT security vulnerabilities	Suggested protocols, firewalls, segmentation	Solutions ensure integrity, confidentiality, availability
18	Nasr Abosata et al.	2021	IoT layer-based classification & literature review	Surveyed attacks & countermeasures for IIoT	Discussed cryptography, IDS, protocols	Offers holistic IIoT security overview
19	Mehmet Bahar et al.	2024	Systemic & cognitive security framework + case study	Military simulation to explore IoT components	Showed trust, privacy, access challenges	Cognitive system design supports secure architecture
20	Mehamed Ahmed Abdurahman	2018	Survey of CIoT trends, technologies, architectures	Explores enabling technologies and open issues	Identified cognitive gaps in IoT solutions	CIoT enhances application-fit and industrial optimization

TABLE 2.2
Significant Contributions

S.No	Authors Name	Year	Simulation Parameter	Value
1	Bharath Reedy Konatham et al.	2023	Accuracy (CNN+GRU)	94.94%
2	T. Simra et al.	2023	Precision (CNN+GRU)	98.49%
3	A. Ghimire et al.	2023	Recall (CNN+GRU)	92.29%
4	F. Amsaad et al.	2023	F1-Score (CNN+GRU)	95.24%
5	M. I. Ibrahem et al.	2023	False Alarm Rate (FAR)	0.001
6	Liming Fang et al.	2020	Performance Evaluation	Low Overhead
7	Liming Fang et al.	2020	Data Access Control	Fine-Grained
8	Anonymous (Blockchain Auth Mechanism)	2022	Security Feature	Enhanced Privacy
9	Anonymous (Blockchain Auth Mechanism)	2022	Attack Mitigation	Successful

A significant contribution of this work was the proposal of a layered security architecture for CIoT, which encompasses device-level, network-level, cognitive engine-level, and cloud-level protections. This holistic model seeks to safeguard the data lifecycle from initial sensing and processing to final storage and analysis. The authors stressed the importance of incorporating privacy-preserving techniques such as homomorphic encryption, federated learning, and differential privacy into the cognitive layer to mitigate risks without compromising functionality. Despite the conceptual richness of the study, the authors acknowledged that their work was primarily theoretical and analytical in nature. There was no empirical validation or simulation-based performance analysis presented in the paper. Nevertheless, the research holds significant value as a foundational exploration into the intersection of cognitive technologies and IoT security. It lays the groundwork for future studies that could build on these theoretical frameworks by integrating them into experimental CIoT prototypes and evaluating their effectiveness in real-world environments. K. Devi and V. Ramesh, in 2023, proposed a novel security framework integrating blockchain technology and ML for safeguarding IoT ecosystems. Their approach focused on leveraging the decentralized nature of blockchain to ensure data integrity, transparency, and secure authentication across distributed IoT nodes. By enabling each device to function as a blockchain node, the system reduced dependence on centralized authorities and mitigated risks related to single points of failure. The authors introduced a multi-layered architecture in which blockchain handled secure transaction logging and smart contract execution, thereby enforcing predefined access controls and triggering automated security responses.

In conjunction with blockchain, the system incorporated ML algorithms trained to identify anomalous patterns in network traffic and device behavior. This dynamic anomaly detection mechanism allowed real-time monitoring and adaptive threat mitigation, making the security solution suitable for rapidly changing IoT environments. Alhough the study lacked detailed experimental or simulation results, the proposed model demonstrated strong theoretical potential, especially in critical application domains such as smart cities, industrial IoT, and healthcare systems. The dual-layer design emphasized not only the robustness of blockchain-based authentication but also the adaptability of intelligent detection systems, forming a comprehensive foundation for future security architectures in IoT networks [9]. G. Sangeetha and A. Selvaraj, in 2022, presented a comprehensive review on the security challenges faced in CIoT applications. Their study examined how the integration of cognitive computing into IoT environments enhances both functionality and complexity, particularly in smart ecosystems like healthcare, industrial automation, and smart homes. The authors outlined multiple security concerns associated with CIoT, including sensor spoofing, data manipulation, adversarial learning, and context-aware attacks. By categorizing these threats across physical, network, and cognitive layers, the paper provided a layered threat model essential for designing resilient security protocols. Furthermore, the review highlighted the importance of adaptive security mechanisms based on cognitive learning, capable of adjusting defense strategies in real-time based on environmental stimuli and user behavior. The authors discussed a variety of proposed countermeasures such as lightweight encryption, distributed trust management, and context-driven anomaly detection. Despite the lack of experimental validation or performance benchmarks, their work served as an essential reference for understanding the dynamic and evolving nature of threats in CIoT systems. The paper underscored the growing need for intelligent, self-learning security solutions that align with the complexity of modern smart applications [10].

C. Rama Mohan Reddy and M. Rupa, in 2022, published "Security Framework for IoT Using Machine Learning Techniques." Their work introduced a flexible and scalable framework tailored for securing IoT environments using both supervised and unsupervised ML methods. The central idea of the framework was to analyze IoT network traffic and classify it into either benign or malicious behavior using attributes such as packet size, communication protocol, and timestamp data. The framework was designed to accommodate a wide range of attack vectors, including known and unknown threats, and was capable of adapting over time by retraining the underlying models as new data became available. This adaptability made the system especially relevant for dynamic and heterogeneous IoT ecosystems. Although the paper did not disclose specific datasets or empirical values, the authors emphasized the system's ability to achieve high detection accuracy across diverse IoT applications. The modular architecture allowed for the seamless integration of new features and detection strategies without overhauling the entire system, enhancing maintainability and scalability. The framework's emphasis on retrainability and behavioral analysis rather than fixed signatures added robustness against zero-day attacks. This contribution significantly advanced the use of intelligent security frameworks for IoT systems by demonstrating the feasibility and practicality of integrating ML into threat detection pipelines, even though it lacked experimental benchmarking [11].

A. M. Bhavatharini et al., in 2022, presented "Smart IoT Security using AI Techniques," a study focused on leveraging AI for proactive and adaptive security in smart IoT networks. The authors explored a diverse set of AI algorithms including Support Vector Machines (SVM), Decision Trees, and Neural Networks, each selected for its unique strength in pattern recognition and classification. Their proposed hybrid model was capable of dynamically updating its security rules based on evolving data streams, thereby improving the system's responsiveness to emerging and sophisticated threats. The AI-driven approach moved beyond conventional rule-based systems by learning from anomalies and adapting policies without requiring manual intervention. Though specific simulation metrics were not reported, the authors claimed notable improvements in threat detection and response times when compared to traditional security mechanisms. The system was designed with smart environments in mind—particularly in use cases like smart homes, smart grids, and industrial automation—where devices are constantly interacting and generating complex behavioral data. The adaptability and learning capacity of the AI models allowed for context-aware threat mitigation and predictive analytics, which are critical for maintaining robust security in real-time. Overall, this study demonstrated the transformative potential of AI in building intelligent and autonomous IoT security systems [12].

R. Rani and N. R. Choudhary, in 2022, published "An Overview of Cognitive IoT: Architecture, Enabling Technologies, and Security." This work presented an in-depth exploration of Cognitive IoT (CIoT), focusing on how cognitive computing enhances the capabilities of traditional IoT systems through intelligent decision-making and context-aware operations. The paper dissected the architecture of CIoT by mapping its components—from sensors and edge devices to cloud infrastructure and cognitive agents. It emphasized enabling technologies such as fog computing for latency reduction, edge AI for local decision-making, and cloud computing for scalable analytics. A detailed discussion on the convergence of these technologies illustrated how they collectively support autonomous, intelligent, and resilient CIoT applications. In terms of security, the authors highlighted major vulnerabilities such as data breaches, impersonation, and denial-of-service attacks that are especially problematic in multi-device cognitive environments. To address these challenges, they proposed a layered security model incorporating data encryption, authentication protocols, and cognitive analytics for real-time threat detection. While the paper did not include experimental validation, it provided conceptual models and strategic guidelines for implementing secure and trustworthy CIoT systems. Their contribution laid foundational insights for future research, particularly in designing adaptive trust frameworks and secure data exchange protocols that align with the fluid and intelligent nature of CIoT environments [13].

In 2021, P. R. M. V. Kishore and K. R. N. Raju proposed an "Enhanced Security Model for IoT Devices Using Machine Learning." The study introduced an innovative approach to enhance the security of IoT devices by leveraging ML techniques. Their model incorporated classification algorithms like Random Forest and k-Nearest Neighbors (k-NN), which were employed to detect anomalies and unauthorized access patterns within IoT environments. These algorithms were trained to recognize normal behavior and flag deviations, thus enabling the identification of potential security threats. While the authors did not provide specific simulation metrics, they reported significant improvements in detection rates, which were achieved without introducing substantial latency. This paper underscored the growing importance and potential of ML in creating scalable, proactive security solutions for IoT networks. The authors also highlighted the increasing need for advanced, adaptive security mechanisms to combat evolving threats in IoT environments [14]. Building upon the need for robust security in IoT systems, Emrah Tomur et al., in 2021, published "SoK: Investigation of Security and Functional Safety in Industrial IoT." Their work analyzed the critical relationship between security and functional safety within Industrial IoT (IIoT) systems, where failure in security mechanisms could lead to disastrous consequences due to the interconnectedness of cyber and physical systems. The authors provided a comprehensive taxonomy of threats specifically targeting critical safety functions within IIoT systems, detailing various attack scenarios. These attacks, which could potentially compromise both digital assets and physical infrastructure, were examined in depth. As part of their research, the paper recommended adopting emerging mobile communication security standards such as 5G, alongside the use of safety-hardened communication protocols. While not directly focusing on cognitive security, this study shed light on risk mitigation strategies crucial to safeguarding critical industrial infrastructures. By emphasizing the need for a holistic view of security that integrates both cyber and physical aspects, the authors contributed valuable insights for improving safety and resilience in smart manufacturing environments [15].

Ulla Gain, in the same year, presented "Applying Frameworks for Cognitive Services in IIoT," which explored how cognitive services enabled by AI and big data could enhance business models and operational transparency within IIoT systems. This paper's primary focus was on the role of cognitive technologies in optimizing the performance and decision-making processes of IIoT applications, rather than directly addressing cognitive security. However, it pointed out that the integration of cognitive services could indirectly strengthen security measures through data-driven insights. By leveraging AI and big data, IIoT systems could analyze vast amounts of operational data to identify patterns, detect anomalies, and make informed decisions that enhance the security posture of industrial networks. The author also emphasized the role of cognitive frameworks in improving business processes, enabling better product lifecycle management, and enhancing service transparency. While cognitive security was not the central theme, this work highlighted the broader security benefits associated with AI-enhanced systems in IIoT settings [16]. In 2024, Farzeen Basith et al. published a book chapter titled "Cognitive IoT Unveiled," which offered an in-depth exploration of the integration of AI and ML into the CIoT. The authors highlighted how these technologies could drive the next-generation of IoT systems, enabling them to become more intelligent, adaptive, and autonomous. One of the core themes of this work was the ethical and responsible use of AI in maintaining data integrity, user privacy, and ensuring ethical decision-making. The chapter also delved into the challenges that arise in the deployment of cognitive systems, particularly around issues such as transparency, fairness, and accountability. These concerns are critical for the widespread adoption of CIoT, especially in domains where decisions based on AI could have significant social and security implications. The paper provided a broader societal perspective on CIoT, articulating how its deployment could impact not just technological progress, but also the ethical landscape of IoT systems in everyday life. The authors called for a more responsible approach to designing and deploying CIoT solutions that balance innovation with societal and security needs [17].

In 2014, Arbia Riahi et al. introduced "A Systemic and Cognitive Approach for IoT Security," an early yet influential contribution to the field of IoT security. The authors proposed a systemic and cognitive model to address the growing security concerns associated with IoT systems. This model was conceptualized as a triangular pyramid, comprising four key components: the person, process, technological ecosystem, and intelligent object. The authors emphasized the interplay between these elements, particularly focusing on the tensions that arise from factors such as trust, privacy, reliability, and safety. By considering the cognitive aspects of security, the paper suggested that understanding and addressing human behaviors, technological limitations, and the environment is crucial in securing IoT systems. Furthermore, the paper advocated for the implementation of policies, auditing mechanisms, and thorough documentation practices to ensure effective governance and accountability in the security of IoT systems. This early work laid the foundation for more holistic and integrative approaches to IoT security [18]. In 2020, Liming Fang et al. proposed a "Secure and Fine-Grained Scheme for Data Security in Industrial IoT Platforms for Smart City," presenting a novel security scheme aimed at safeguarding data in Industrial IoT (IIoT) environments within smart cities. Their proposed scheme relied on the conditional proxy re-encryption primitive, which allowed for secure data sharing without compromising privacy or system functionality. The authors demonstrated how this approach could prevent unauthorized access to sensitive data while ensuring that IIoT platforms maintained their functionality across different applications. One of the key benefits of this scheme was its low computational overhead, which made it particularly suitable for data-intensive environments where performance and scalability are crucial. The flexibility of the approach was also highlighted, as it could be adapted to different use cases within smart city infrastructures. This work contributed to the development of practical, efficient solutions to protect data in increasingly complex IIoT environments, offering a balance between security and operational effectiveness [19].

Bharath Reedy Konatham et al., in 2023, presented "ML-assisted Security for Anomaly Detection in Industrial IoT Applications," where they tested a range of deep learning models, including CNN, GRU, Long Short-Term Memory networks (LSTM), and a hybrid CNN+GRU model for anomaly detection in IIoT systems [5]. The results of their study were impressive, with the hybrid CNN+GRU model achieving 94.94% accuracy, 98.49% precision, 92.29% recall, 95.24% F1-score, and a remarkably low false alarm rate of 0.001. Despite the hybrid model having longer convergence times compared to other approaches, its performance in accurately identifying anomalies was outstanding. This study demonstrated the effectiveness of hybrid deep learning models in enhancing the security of IIoT systems by improving anomaly detection capabilities. The authors emphasized the potential of ML in automating and refining security measures in industrial environments, where rapid and accurate threat detection is critical [20]. In 2022, an anonymous paper titled "Industrial IoT with Secure Authentication Mechanism Through Blockchain Technology" introduced a hybrid authentication model designed to enhance the security of Industrial IoT (IIoT) systems. The model combined ECC and PKI with blockchain technology to secure authentication processes and ensure data privacy. By employing private blockchains, the authors aimed to mitigate common attacks such as data leakage, eavesdropping, and replay attacks, which are frequent in IIoT environments. The use of blockchain technology provided a decentralized and immutable ledger for secure transaction tracking, while ECC and PKI ensured that user authentication was both secure and efficient. The mechanism was validated successfully, with the results showing significant promise in enhancing the overall security posture of IIoT systems. This work underscored the potential of blockchain as a transformative technology for securing critical infrastructure in industrial environments [21].

In 2021, Lehlogonolo P. I. Ledwaba and Gerhard P. Hancke explored the "Security Challenges for Industrial IoT," a paper that focused on the security implications of exposing safety-critical industrial operations to external networks. With the increasing interconnectivity of IIoT systems, the authors raised concerns about the vulnerability of industrial processes to cyber-attacks, which could lead to catastrophic consequences. The paper highlighted the need for identifying and implementing appropriate security services across the entire IIoT stack, from device-level security to network and data protection. By focusing on the specific security needs of Industry 4.0, the authors provided a structured approach to developing robust security strategies tailored to the evolving nature of IIoT environments. Their work stressed the importance of integrating security into every aspect of industrial IoT systems to ensure the safety and resilience of critical infrastructure [22]. In 2023, Andrea Sánchez-Zumba and Diego Ávila-Pesantez published "Cybersecurity for Industrial IoT: Threats, Vulnerabilities, and Solutions," a comprehensive study that reviewed 43 papers published between 2017 and 2022. The authors applied the STRIDE model—a widely used framework for threat modelling to classify and analyze the various security threats in Industrial IoT (IIoT) environments. The study highlighted key vulnerabilities and proposed practical solutions to mitigate these risks, including the use of secure communication protocols, firewalls, IDS, and network segmentation. The review provided an in-depth overview of the evolving cybersecurity landscape in IIoT and offered actionable guidelines for securing industrial systems against a broad range of cyber threats. By addressing both technical and procedural aspects of security, the authors contributed valuable insights into safeguarding IIoT systems from increasing and more sophisticated attacks. The paper served as a useful resource for both practitioners and researchers seeking to strengthen the cybersecurity posture of industrial infrastructures [5].

Nasr Abosata et al., in 2021, presented "Internet of Things for System Integrity: A Comprehensive Survey on Security, Attacks and Countermeasures for Industrial Applications," a thorough survey aimed at connecting attack types with the architectural layers of IoT systems. The paper reviewed a variety of security threats facing IIoT systems, such as data breaches, denial-of-service attacks, and unauthorized access, and linked them to the respective layers of IoT architecture. The authors also provided an extensive review of countermeasures such as cryptographic techniques, networking protocols, and IDS, which can be deployed to mitigate these threats. Moreover, the paper discussed emerging tools and frameworks for testing and assessing IoT security, making it an essential reference for researchers working in the field. By synthesizing a wide range of solutions and attack scenarios, this survey offered valuable guidance for improving the security of industrial IoT applications, highlighting the need for both preventive and detective measures in building resilient systems [23].

In 2024, Mehmet Bahar et al. published "Advancing IoT Security: A Systemic and Cognitive Framework for Military Application," which introduced a novel systemic and cognitive security framework tailored for military simulations. This paper examined the complexities of securing IoT systems in high-stakes environments, such as military applications, where the implications of security breaches are particularly severe. The authors illustrated how various system components interact and highlighted the challenges of ensuring privacy, trust, and secure identification within military IoT networks. The framework proposed a holistic view of IoT security, incorporating both technical and cognitive aspects to address the unique needs of military environments. This work provided a roadmap for researchers and practitioners in military IoT security, offering practical guidelines for developing robust security mechanisms that balance functionality, resilience, and strategic security objectives [24]. In 2018, Mehamed Ahmed Abdurahman presented "A Survey on the Concepts, Trends, Enabling Technologies, Architectures, Challenges and Open Issues in Cognitive IoT-Based Smart Environments." This early study laid the groundwork for the field of Cognitive IoT, focusing on its application in smart environments. The paper provided an extensive overview of the enabling technologies, architectural frameworks, and trends shaping the development of Cognitive IoT systems. It discussed the critical role of cognitive intelligence in bridging the gap between the capabilities of IoT devices and the specific requirements of the applications they serve. The paper also addressed the challenges faced by researchers and practitioners, including issues related to data management, real-time processing, and the integration of diverse IoT devices into a cohesive, intelligent system. By offering a comprehensive survey of early-stage research, this work played a pivotal role in shaping future directions for Cognitive IoT and its potential applications in smart cities, healthcare, and other domains [25].

Bharath Reedy Konatham et al. [5] used a hybrid CNN+GRU model for anomaly detection in IIoT, achieving 94.94% accuracy, 98.49% precision, and 92.29% recall, which indicates the model's high reliability in detecting true anomalies while minimizing false positives. The F1-score of 95.24% reflects a balanced performance, efficiently capturing most anomalies without triggering unnecessary alarms. Additionally, the model's extremely low false alarm rate (0.001) demonstrates its precision in minimizing disruptions in IIoT systems, making it highly effective for real-time anomaly detection. Liming Fang et al. (2020) introduced a security scheme for IIoT in smart cities, emphasizing low computational overhead and fine-grained data access control. By using conditional proxy re-encryption, it ensured secure data sharing and prevented unauthorized access, maintaining system functionality while safeguarding sensitive information. The scheme's low overhead and flexible access control made it an ideal solution for scalable, data-intensive IIoT environments. In 2022, an anonymous study proposed a hybrid authentication mechanism combining ECC, PKI, and blockchain for IIoT security. This approach enhanced privacy by utilizing private blockchains and successfully mitigated common attacks, such as data leakage, eavesdropping, and replay attacks, demonstrating the potential of blockchain in securing industrial IoT environments.

2.3 DATA PRIVACY AND ETHICS IN COGNITIVE SYSTEMS

As cognitive security systems in Industrial IoT (IIoT) increasingly rely on vast volumes of real-time data and AI-driven analytics, ensuring data privacy becomes a critical challenge. Cognitive systems often require access to sensitive operational and personal information to make informed security decisions. However, this raises concerns about user consent, data ownership, and ethical boundaries in data utilization. While ML and deep learning offer intelligent threat detection, they can also introduce bias, opacity, and lack of explainability, making it difficult to validate their decisions or ensure fair use. Moreover, ethical frameworks for responsible AI usage—especially in autonomous security responses—are still underdeveloped. The integration of privacy-preserving techniques like differential privacy and federated learning is still in its infancy in IIoT contexts, indicating a substantial research gap in creating transparent and ethically aligned cognitive security models.

2.4 RESOURCE CONSTRAINTS IN INDUSTRIAL DEVICES

IIoT devices, particularly edge nodes and embedded sensors, often operate under strict resource limitations in terms of processing power, memory, energy, and bandwidth. Implementing cognitive security algorithms—especially those that involve deep learning or complex encryption schemes—can be computationally intensive and incompatible with the lightweight nature of many industrial devices. While studies such as those using hybrid ML models (e.g., CNN+GRU) have demonstrated high accuracy in threat detection, their deployment on constrained hardware can lead to latency issues or system instability. Additionally, real-time data analysis and decision-making often require offloading computations to the cloud or edge, introducing potential security risks and latency. Addressing this challenge requires the development of lightweight, energy-efficient cognitive algorithms specifically designed for constrained IIoT environments—an area where further research is urgently needed.

2.5 LACK OF STANDARDIZATION AND INTEROPERABILITY

The heterogeneous nature of IIoT systems, comprising diverse devices, communication protocols, and security frameworks, leads to significant challenges in standardization and interoperability. Cognitive security solutions must be adaptable across various platforms, yet current implementations are often designed in siloed environments without considering cross-system compatibility. This fragmentation hampers the development of unified security architectures that can function cohesively across different industrial setups. Moreover, the absence of standardized data formats and communication protocols complicates the integration of AI-driven cognitive systems into existing IIoT infrastructures. Research must focus on developing interoperable cognitive security frameworks and global standards that facilitate seamless communication, policy enforcement, and threat detection across heterogeneous IIoT ecosystems.

2.6 CONCLUSION

The security of IoT systems, particularly in industrial and smart city contexts, is a growing concern that requires multifaceted solutions. The integration of ML models, such as the hybrid CNN+GRU model, shows promising results in anomaly detection, with high accuracy and minimal false positives, which are crucial for the operational reliability of IIoT systems. Furthermore, solutions like the low-overhead security scheme proposed by Liming Fang et al. (2020) and the blockchain-based authentication mechanism introduced in 2022 demonstrate the potential of advanced encryption and decentralized technologies in enhancing privacy and preventing attacks in IoT environments. Collectively, these studies underscore the importance of a holistic approach to IoT security, one that combines advanced ML techniques with encryption and blockchain technologies. As IoT systems continue to evolve, these solutions provide a strong foundation for building secure, efficient, and resilient IoT networks. Future research should continue to refine these methods and explore their integration across diverse IoT applications to ensure robust security against an ever-growing landscape of cyber threats.

REFERENCES

[1] Loupos, K., Caglayan, B., Papageorgiou, A., Starynkevitch, B., Vedrine, F., Skoufis, C., Christofi, S., Karakostas, B., Mygiakis, A., Theofilis, G., Chiappetta, A., Avgoustidis, H., & Boulougouris, G. (2019). Cognition Enabled IoT Platform for Industrial IoT Safety, Security and Privacy—The CHARIOT Project. *Computer Aided Modeling and Design of Communication Links and Networks*, 1–4. https://doi.org/10.1109/CAMAD.2019.8858488.

[2] Zheng, W., Lai, C.-F., He, D., Kumar, N., & Chen, B. (2021). Secure Storage Auditing with Efficient Key Updates for Cognitive Industrial IoT Environment. *IEEE Transactions on Industrial Informatics*, 17(6), 4238–4247. https://doi.org/10.1109/TII.2020.2991204.

[3] Jha, S. (2021). Improved Powered Cognitive Sensor Nodes Proposed WSN for a Heterogeneous CR-WSNs. 2021 Asia-Pacific Conference on Communications Technology and Computer Science (ACCTCS) (pp. 123–124). Shenyang, China. https://doi.org/10.1109/ACCTCS52002.2021.00033.

[4] Sánchez-Zumba, A., & Ávila-Pesantez, D. (2023). Cybersecurity for Industrial IoT, Threats, Vulnerabilities, and Solutions: In X. S. Yang, R. S. Sherratt, N. Dey, & A. Joshi (Eds.), *Proceedings of Eighth International Congress on Information and Communication Technology (ICICT 2023)*, Lecture Notes in Networks and Systems, vol. 693 (pp. 1101–1112). Springer International Publishing. https://doi.org/10.1007/978-981-99-3243-6_90.

[5] Konatham, B. R., Simra, T., Ghimire, A., Amsaad, F., Ibrahem, M. I., & Jhanjhi, N. Z. (2023). ML-assisted Security for Anomaly Detection in Industrial IoT (IIoT) Applications. 1120–1127. https://doi.org/10.1109/smarttechcon57526.2023.10391331.

[6] Abosata, N., Al-Rubaye, S., Inalhan, G., & Emmanouilidis, C. (2021). Internet of Things for System Integrity: A Comprehensive Survey on Security, Attacks and

Countermeasures for Industrial Applications. *Sensors*, 21(11), 3654. https://doi.org/10.3390/S21113654.

[7] Bahar, M., Pattnaik, O., & Baldhare, P. (2024). Advancing IOT Security: A Systemic and Cognitive Framework for Military Application. *International Journal of Innovative Science and Research Technology*, 1246–1250. https://doi.org/10.38124/ijisrt/ijisrt24jun677.

[8] Abdurahman, M. A. (2018). A Survey on the Concepts, Trends, Enabling Technologies, Architectures, Challenges and Open Issues in Cognitive IoT Based Smart Environments. *International Journal of Scientific Research in Science, Engineering and Technology*, 4(1), 512–522. https://doi.org/10.32628/IJSRSET 1841111.

[9] Jha, S., Ahmad, S., Abdeljaber, H. A. M., Othman, N. A., Nazeer, J., Yadav, K., & Alkhayyat, A. (2024). Enabling Resilient Wireless Networks: OFDMA-Based Algorithm for Enhanced Survivability and Privacy in 6G IoT Environments. *IEEE Transactions on Consumer Electronics*. https://doi.org/10.1109/TCE.2024.3370414.

[10] Chen, H.-C., Chen, H.-C., Yang, W., & Chou, C.-L. (2021). An Online Cognitive Authentication and Trust Evaluation Application Programming Interface for Cognitive Security Gateway Based on Distributed Massive Internet of Things Network. *Concurrency and Computation: Practice and Experience*, 33(19). https://doi.org/10.1002/CPE.6128.

[11] Jha, S., Jha, N., Prashar, D., Ahmad, S., Alouffi, B., & Alharbi, A. (2022). Integrated IoT-Based Secure and Efficient Key Management Framework Using Hashgraphs for Autonomous Vehicles to Ensure Road Safety. *Sensors*, 22(7), 2529. https://doi.org/10.3390/s22072529.

[12] Chen, H., Hu, M., Yan, H., & Yu, P. (2019). Research on Industrial Internet of Things Security Architecture and Protection Strategy. *International Conference on Virtual Reality*. https://doi.org/10.1109/ICVRIS.2019.00095.

[13] Kumar, K. B. N., & Rao, M. P. (2017). A Novel Cognitive Security Approach for Internet of Things. *International Journal of Engineering and Technology*, 9, 579–584. https://doi.org/10.21817/IJET/2017/V9I3/170903S087.

[14] Portal, G., de Matos, E., & Hessel, F. (2020). An Edge Decentralized Security Architecture for Industrial IoT Applications. *The Internet of Things*, 1–6. https://doi.org/10.1109/WF-IOT48130.2020.9221176.

[15] Tomur, E., Gulen, U., Ustundag Soykan, E., Ersoy, M. A., Karakoç, F., Karacay, L., & Comak, P. (2021). SoK: Investigation of Security and Functional Safety in Industrial IoT. *Computer Science Symposium in Russia*, 226–233. https://doi.org/10.1109/CSR51186.2021.9527921.

[16] Gain, U. (2021). Applying Frameworks for Cognitive Services in IIoT. *Journal of Systems Science and Systems Engineering*, 30(1), 59–84. https://doi.org/10.1007/S11518-021-5480-X.

[17] Basith, F., Deepti, A. R., Vivek, K. S., & Manimekala, B. (2024). Cognitive IoT Unveiled. In Ivan Giannoccaro (Ed.), *Advances in Computational Intelligence and Robotics Book Series*, 1–32. IGI Global.

[18] Jha, S., Prashar, D., Elngar, A. A. (2020). A Novel Approach Using Modified Filtering Algorithm (MFA) for Effective Completion of Cloud Tasks. *Journal of Intelligent & Fuzzy Systems*, 39(6), 8409–8417. https://doi.org/10.3233/JIFS-189159

[19] Aghila Rajagopal, S. A., Jha, S., Abdeljaber, H. A., & Nazeer, J. (2023). AI Based Secure Analytics of Clinical Data in Cloud Environment: Towards Smart Cities and Healthcare. *Journal of Advances in Information Technology*, 14(5), 1132–1142.

[20] Jha, S., Sultan, A., Alharbi, M., Alouffi, B., & Sebastian, S. (2021). Secured and Provisioned Access Authentication Using Subscribed User Identity in Federated Clouds. *International Journal of Advanced Computer Science and Applications*, 12(11), 178–187.

[21] Riahi, A., Natalizio, E., Challal, Y., Mitton, N., & Iera, A. (2014). A Systemic and Cognitive Approach for IoT Security. In *Proceedings of the 2014 International Conference on Computing, Networking and Communications* (ICNC 2014) (pp. 183–188). IEEE. https://doi.org/10.1109/ICCNC.2014.6785328.

[22] Fang, L., Zhang, H., Li, M., Ge, C., Liu, L., & Liu, Z. (2020). A Secure and Fine-Grained Scheme for Data Security in Industrial IoT Platforms for Smart City. *IEEE Internet of Things Journal*, 7(9), 7982–7990. https://doi.org/10.1109/JIOT.2020.2996664.

[23] Jha, S., Ahmad, S., Alkhayyat H. A. M., Othman, N. A., Nazeer, J., Yadav, K., & Alkhayyat, A. (2024). Enabling Resilient Wireless Networks: OFDMA-Based Algorithm for Enhanced Survivability and Privacy in 6G IoT Environments. *IEEE Transactions on Consumer Electronics*, 70(1), 3810–3819. https://doi.org/10.1109/TCE.2024.3370414.

[24] Jha, S., Prashar, D., Long, H. V., & Taniar, D. (2020). Recurrent Neural Network for Detecting Malware. *Computers & Security*, 99, 102037. https://doi.org/10.1016/j.cose.2020.102037.

[25] Jha, S., Nkenyereye, L., Joshi, G. P., & Yang, E. (2020). Mitigating and Monitoring Smart City Using Internet of Things. *CMC-Computers, Materials & Continua*, 65(2), 1059–1079. https://doi.org/10.32604/cmc.2020.011754.

3 Advancing Security in Industry 4.0

Ritu Raj Choudhary, Udit Mamodiya, Prerna Srivastava, and Sultan Ahmad

3.1 INTRODUCTION

The onset of Industry 4.0 has transformed the industry scene by bringing together cyber-physical systems, automation under an AI microscope, and the Industrial Internet of Things (IIoT). In this paradigm shift, there is dire necessity of redefining how we can conceptualize and actualize cybersecurity. Since information transfers exponentially among devices, networks, and autonomous agents, perimeter-based security modus operandi is no longer viable. Rather, cognitive security systems, which are intelligent, adaptive, and self-learning, appear as critical mechanisms of defense against industrial infrastructures when faced with forms of an ever more advanced and resilient threat. The IIoT age requires their security solutions to be able to think, adapt, and learn according to Prof. Maurice Dawson, as his article explains: the age of the IIoT requires on the one hand systems and technologies capable of defense that can also think, adapt, and learn just like the adversary that they protect. Using machine learning (ML), behavioral analytics, and real-time threat intelligence, cognitive security has the capability to not only detect anomalies, predict intrusions, and carry out cyber-incident response in real-time, without the involvement of a human operator. It is not only a technical change of paradigm but rather a philosophical change. It suggests to re-think trust, transparency and responsibility within the governance of intelligent systems. Nonetheless, although much is said about the technical performance of such systems [1], there is a considerable socio-ethical lag in the practice. The question of interpretability, equity, data protection, and compliance has emerged as critical problems. Floridi et al. state that the more we entrust decision-making to machines, the more we will need to make sure that the decisions made and actions taken by them are equal to the values the society holds [2]. In critical infrastructures (where not only data, but lives, are at stake) this alignment is a non-issue.

Although initiatives such as those of UNESCO and the European Commission suggest the introduction of the ethical framework of AI, its adoption by industrial sector stakeholders is still poor and uneven [3, 4]. There are many industries that continue to lack the rule of law in governance AI models and, in certain instances, even a form of accountability. As best demonstrated by the World Economic Forum, recently stating, "the security blind spot and ethical dilemmat that the absence of

DOI: 10.1201/9781003609773-3

monitoring and diagnostic surveillance persist [11]. Hospitals face the dual challenge of deploying robust AI-based defence mechanisms while upholding strict confidentiality and compliance standards.

3.4 CORE PRINCIPLES FOR THE FUTURE OF IIOT SECURITY

Securing the future of Industry 4.0 requires a shift from fragmented and reactive approaches to holistic, proactive, and ethically aligned security paradigms. As cyber-physical systems become more intelligent and interconnected, Industrial IoT (IIoT) environments demand core security principles that are not only robust and intelligent but also transparent, inclusive, and aligned with societal values.

Table 3.2 operationalizes several core concepts in cognitive security by relating them to concrete threats in reality and how it can be implemented through certain deployment strategies within industry.

3.4.1 Security Proactive versus Security Reactive

Traditional industrial cybersecurity has been reactive to a great extent and takes place after threats occur. But, to those threats that are more modern (and allowed to operate to be more autonomous and time-sensitive), the anticipatory defense mechanisms are needed. The proactive security puts a strong focus on predicting threats, real-time analytics, and independent remediation [1, 16]. The use of cognitive systems that are trained using the behavioral and situation data will help industries to spot the preattack signs and eliminate the threat before it develops further. Dynamic systems and continuous threat modeling opportunity and even dynamic policy adaptation is also available in proactive systems that allows to be prepared to new and emerging brain deception vectors.

TABLE 3.2
Core Principles for Secure IIoT—Alignment with Risk Mitigation Goals

Security Principle	Definition	Risk Mitigated	Implementation Strategy
Explainability	Making AI decisions transparent and traceable	Operator distrust, model misclassification	Use of XAI tools like LIME, SHAP; interpretable ML models
Fairness	Ensuring algorithms are free of systemic bias	Workforce alienation, skewed results	Auditing data inputs; fairness constraints in ML pipelines
Privacy-by-Design	Embedding privacy into system architecture from inception	Unauthorized data access, compliance breach	Data minimization, encryption, federated-learning protocols
Human Oversight	Retaining decision authority or intervention capabilities	Over-automation errors, ethical ambiguity	Human-in-the-loop (HITL) & human-on-the-loop (HOTL) governance

3.4.2 Incorporating Security Design of Explainability, Fairness and Privacy

Cognitive security of the future depends on the ability to construct systems that are intelligent, explainable, and defensible. The provisions of XAI should be implemented in security models to make any alerts and actions traceable and detectable by the human operator [9, 10]. Besides, fairness of algorithms is highly important to avoid the problem of prejudiced detection of threats, which can negatively affect particular workers, behaviors, or devices. When the data used to train AI systems are biased or incomplete, there is a high probability that they would reproduce previously existing inequality or false positive. Privacy-by-design is just as significant, this means that data minimization, consent process and encryption need to be part of the design system, rather than value-added. Since in IIoT, both personal and operational data are always created, privacy should be respected not only ethically but also on regulatory grounds [3] embedded IIoT. In this context, the ethical concept can benefit the security of the IIoT framework by managing risks and limiting the potential hazards associated with security breaches.

The infographic Figure 3.1 depicts a layered architecture in which principles of core ethical ethos, namely, transparency, fairness, privacy and human oversight, are embedded into the functional design of Industrial AI systems. It points out the connection in ethical governance to IIoT security layer of the data acquisition, intelligent threat analysis, decision execution and policy enforcement.

3.4.3 The Necessity of Human Oversight and Human–AI Collaboration

The complexity of IIoT security should never make human attention replaceable as the AI makes the decisions. Both human-in-the-loop (HITL) and human-on-the-loop

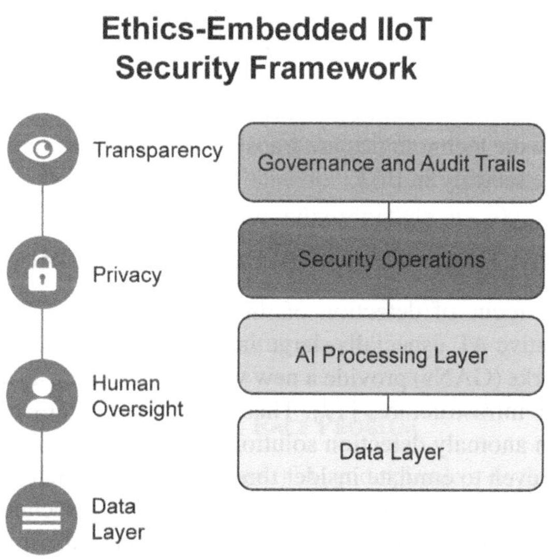

FIGURE 3.1 Security framework of ethics.

hardening and retraining and red-teaming. It needs some new methods of defence against these adversarial threats, including effective training, sanitizing input, uncertainty quantification, and ensembles of multiple layers of models. The industrial cybersecurity capabilities and threats are swiftly changing due to emerging technologies. In order to reap the benefits of the same, organizations have to engage in research-based implementation, architectures that are ready-to-govern and preseers. It is still the unknown future, that will be enjoyed by individuals after having embraced such technologies, but have properly embraced them with responsibility and with vision.

3.5.6 Strategic Framework: A guide to Industry 4.0 security

With cognitive security and the development of Industry 4.0, organizations are urgently in need of a shift toward a strategic wholesale security deployment, as the current tendency is toward isolated or reactive deployments of defence. This roadmap cannot just be a passive adoption of the technology but rather penalty on ethical justifications, workforce, and the necessary governance flexible enough to adapt in the long run. In order to establish this transformation, the three phases are offered below: short-term objectives (2025–2027), mid-term integration objectives (2028–2030), and the long-term vision governance of autonomous and ethically controlled security ecosystems to be achieved beyond 2030.

3.5.6.1 Phase I: Short Term Priorities (2025–2027)

The first phase is characterized by the focus on the establishment of the groundwork of the secure cognitive integration within the IIoT systems. To start encouraging internal AI governance structures, organizations should fill in the infrastructural gaps and start formalizing their AI governance structures. The essential measures during this step consist of implementing AI governance through the creation of ethics boards, compliance officers, and risk assessment units according to international best practices like ISO/IEC and NIST. It is also important to implement the tools for explainability and transparency like visual dashboards and interpretable models that improve trust in operators and limit their resistance [8, 10]. The other pillar is the implementation of zero-trust architecture which replaces the earlier reliance on perimeter security with micro-perimeterization and least-privileged access mechanism. Also, companies must engage in work-related training programs and offer AI security literacy initiatives to engineers, operators, and managerial personnel. The milestone of this phase will be to be able to set an actualized ethical and working normal of cognitive security in each of the primary critical IIoT endpoints.

Figure 3.2 shows a radar chart showing five core capabilities of explainability, real-time detection, fairness, interoperability, and resilience against five key industrial risk areas to demonstrate the level of security maturity. The chart shows that real-time detection and resilience are strong, whereas fairness and interoperability become aspects that require high strategic focus.

3.5.6.2 Phase II: Medium the Term Objectives (2028–2030)

The second phase is about scaling secure and ethical intelligence in the operations of the organization. It is a time that highlights integration between systems, cross-functional

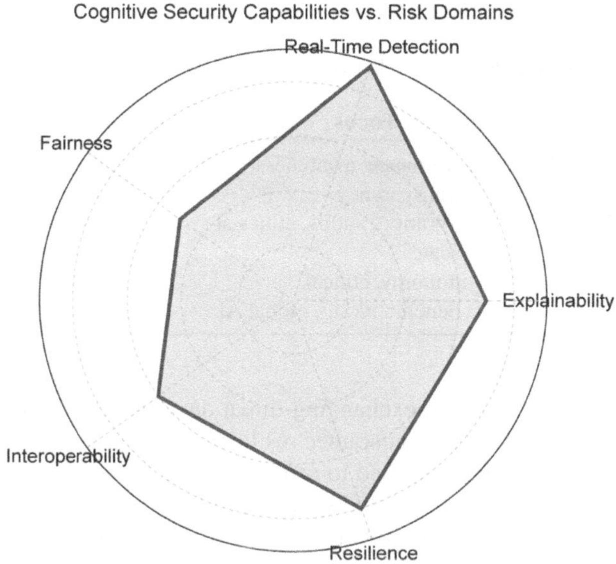

FIGURE 3.2 Radar chart of cognitive security capabilities versus risk domains.

installation and ethical institutionalization. Some of the most critical measures are the implementation of federated security intelligence using cooperative AI models that do not tamper with data privacy and do not violate regional data regulations (Li et al., 2020). By making all AI security tools compatible with such stable criteria as FAIR (Findable, Accessible, Interoperable, Reusable) and FAT (Fairness, Accountability, Transparency), organizations will have to institutionalize ethics-by-design as well. Moreover, companies must engage in AI-powered alternative audit and red teaming which acts like simulated hacking attacks and allows checking the system strength on a regular basis. In order to realize scaleable interoperability, organizations are advised to embrace and develop universal protocols that have global acceptance like OPC UA and IEC 62443. The benchmark of this step is to create a safe, explainable, and interoperable AI infrastructure which follow both local and international rule of cyber security.

3.5.6.3 Phase III Long-Term Vision (after 2030)

The last stage of the strategical roadmap is the appearance of autonomous, ethical, and self-adaptive industrial defense infrastructures. At this point, therefore, the cognitive security would run with minimal human supervision, be capable of adapting to and learning the environment, protecting against new threats, and growing back via federated and swarm-based intelligence. Organizations will have to mount an autonomous security agent, that is, a self-healing, self-defending node, which is real-time adaptively reconfigurable. At this point ethical maturity takes center stage and companies adopt governance benchmarking model this has seen the emergence of companies adopting legal benchmarking models like the Cognitive Security Ethical Maturity Model (CSEMM). In addition, the world needs to join AI-led cybersecurity

TABLE 3.3
Three-Phase Strategic Framework for Responsible AI Evolution

Phase	Timeframe	Primary Focus	Outcome
Phase I	2025–2027	Governance, awareness, transparency	Ethical and operational baseline
Phase II	2028–2030	Integration, audits, ethics at scale	Secure and interoperable cognitive frameworks
Phase III	2030 onward	Autonomy, ethical benchmarking, global AI	Trusted, adaptive, and resilient ecosystems

coallitions and this will help in exchanging threat data in a cross-border format in a very secure manner. As a way to guarantee resilience to the point of live deployment, digital twin simulations must be used to test and tune the security strategies prior to real-world deployment in both real and synthetic environments. The final goal of this stage is to realize trusted autonomy, that is, cognitive security systems are transparent, ethically upright, and strategically aligned with human and organizational values.

Table 3.3 organizes a long-term vision for AI development and governance into three distinct phases, each defined by a specific timeframe, primary focus, and desired outcome. The phased approach ensures that foundational elements are established early, scalable mechanisms are developed in the middle term, and future-forward strategies are implemented in the long run.

3.6 IMPERATIVES IN POLICY AND COLLABORATION

Technological excellence is not the only thing that can ensure a future in the industry 4.0. The further development of cognitive security systems in terms of complexity and autonomy raises concerns about the safe and ethical application of such systems, the pinnacle of which is the development of high-quality policy frameworks, intersectoral collaboration, and international coordination. This section contains the necessary policy considerations and institutional collaborations that must be undertaken to establish a responsible, resilient, and inclusive industrial cybersecurity ecosystem.

3.6.1 Coordinating International Rules

The lack of coherent AI and cybersecurity regulation in different countries is a major stumbling block to deployment of secure and interoperable AI and cybersecurity. Although the EU (2024) has introduced the novelty of a risk-based approach to the regulation of the AI system that is of high risk, including critical infrastructure, other regions lack legal specifics [4]. This regulatory asymmetry causes regulatory arbitrage in that companies can move to those jurisdictions with the least regulations and in effect puts systems under unnecessary risks. There should be a focus on the harmonization of regulations that focus on AI security, privacy, and transparency at the cross-border scale. Multilateral policy dialogues should be led by such institutions as the OECD, ITU, and UNESCO. Such initiatives are also expected to take into consideration the industry-specific standards like IEC 62443 (industrial cybersecurity), ISO/IEC 27001 (information security management).

3.6.2 Ethical Innovation Multi-Stakeholder Collaboration

Governments, industries, academia, and civil society need to escape the silo nature of their operations and collaborate in the co-creation of governance systems that demonstrate both innovation and accountability. Brundage et al. (2023) argue that a strong AI governance should involve engineers, policy professionals, ethicists, and communities that are impacted by AI. This makes the solution in terms of cognitive security technically viable, legal, and socially acceptable.

Some examples of this sort of cooperation are:

- Public-Private partnerships (PPPs) on secure-by-design research
- The collaboration of industry-academia to test and corroborate ethical frameworks.
- AI ethics labs and ethics halls to audit benchmark algorithmic decision making
- Threat intelligence exchange cybersecurity information-sharing platforms

Such mutual models also assist in correcting the aspects of power inequalities because smaller corporations and underrepresented groups would have said in what comes out of the technological creation that affects them.

3.7 INTERNATIONAL THREAT INFORMATION AND INFORMATION EXCHANGE

The current cyber threats are transnational, which are found at the industry 4.0 era. Jurisdictional loopholes, data localization regulations, and a lack of international collaboration are a common source of frustrations, which are used by the attackers. Hence, the governments and organizations should be devoted to common threat intelligence structures. The Global Forum on Cyber Expertise (GFCE) and FIRST (Forum of Incident Response and Security Teams) are some of the initiatives used to encourage the safe cooperation.

On the industry level, the industry manufacturers are supposed to combine with the cybersecurity expertise centers and must have data governance practices that are correlated to both GDPR-like privacy approaches as well as with NIST guidelines of how critical infrastructure is to be safeguarded.

3.8 POLICY LITERACY AND CAPACITY BUILDING

Finally, education and capacity building are a requirement for policy and collaboration performance in the long-term. When the policymaker keeps updated on the latest development in technology points, and the technologists are legally and ethically trained, a policy is made. By introducing cross-disciplinary educational pathways in AI ethics, cybersecurity law, and risk governance, researchers will form a generation of leaders who will be able to advance in this rapidly growing socio-technical landscape [3].

Table 3.4 breaks down a long-term vision of AI development and governance into three phases that have been identified by the key time frame, major emphasis, and desired result. The staggered method will allow establishing the building blocks in the phase of initial development of the platform, building the mechanisms that can be scaled in the medium term, and incorporating the lenses toward the future in the long term. Having internalized policy foresight and collaborative governance in the core of cybersecurity strategy, the stakeholders of the Industry 4.0 can ensure that the advancement of technologies will not contradict the levels of societal trust, human rights, and international safety.

3.9 KEY FINDINGS

Developing Security in Industry 4.0 Far-sighted prospects of the changing nature of cybersecurity in the context of Industry 4.0 are clearly outlined in this chapter. The most important finding is that cognitive security, composed of AI, ML, and human control, has become a crucial support to defend industrial cyber-physical structures. This chapter features a very serious paradigm change in relation to reactive perimeter security that lacks intelligence, proactive security ecosystems that anticipate, identify, and automatically respond to the sophisticated attacks in real-time. The third key product of this chapter is the creation of a three-phase strategic roadmap to apply cognitive security to the sectors. This includes:

- Short-term (2025–2027): Setting the grounds by using the conditions of AI governance, visibility, and zero-trust architectures.
- Mid-term (2028–2030): Scaling of interoperable and ethical AI systems via federated learning, uniform qualities of a protocol, as well as internal reviews.

Developing self-functioning, morally capable, self-healing security infrastructure. This chapter includes the specific results in the sector, demonstrating the prompts of manufacturing, logistics, energy, and healthcare industries that use AI to enhance its orientation through predictive maintenance and real-time detection of threats. Nonetheless, it also brings out such risks as: Reliance on low-explainable black-box AI obscurity Opposition of the employees because of privacy and job security issues

TABLE 3.4
Summary of Policy and Collaboration Priorities

Priority Area	Key Action
Regulatory Harmonization	Align national frameworks; adopt risk-based AI policies
Stakeholder Collaboration	Support public–private–academic coalitions
Cross-Border Intelligence	Establish international data-sharing and joint-response mechanisms
Capacity Development	Build institutional literacy in AI ethics and cybersecurity law

Non-compliance resulting due to poor data governance The second vital lesson is that of making it part of the AI security tools development and implementation to incorporate ethical considerations namely, explainability, fairness, privacy-by-design, and human-in-the-loop oversight. This is consistent with the wider production of this chapter: a proposal of socio-technical framework that incorporates innovation along with the human values and international regulation. Lastly, this chapter emphasizes the need to harmonize the international policies and collaboration across stakeholders to find solutions to global, cross-border connection of cyber threat in the Industry 4.0 era. Among the things it recommends is sharing of threat intelligence across borders, capacity building on AI ethics, and adoption of internationally recognized standards to guarantee trust, safety, and resilience.

3.10 CONCLUSION

With Industry 4.0 transforming the industrial environment through hyperconnectivity and automation and data-driven intelligence, the concept of cognitive security has gained prominence and currently is a main pillar on which the security of the critical infrastructure can be based. In this chapter, we have discussed the potential of AI-enabled security layers mega shift, as well as the ongoing problem of ethical alignment and transparency, and trust. Although cognitive systems have greatly transformed threat detection and predictive maintenance, the experience of their implementation has not been error-free, such as high risks of black box and privacy, not to mention regulatory discrepancy and staff rejection. These predicaments bring out one key lesson, there should be a balance of technological superiority with moral toughness. There is a need to embrace a socio-technical approach that combines smart automation and human supervision, ethical vision, and policy alignment to develop a secure future of Industry 4.0. This should be combined with entrenching explainability, fairness, and privacy in the design of these systems such that AI operates efficiently, transparently, and justly.

With hyperconnectivity, automation, and data-driven intelligence being the hallmarks of the technological transformation that the Industry 4.0 involves in transforming industrial sectors, cognitive security has taken the spot to become a pillar of securing critical infrastructure. The combination of AI with blockchain technology can help make IIoT more resilient as it offers both transparency and accountability and introduces a level of decentralized trust. However, shifting to this paradigm introduces new vulnerabilities, which include black-box opacity, data privacy risks, regulatory fragmentation, and workforce skepticism, that need to be handled preventively. In order to resolve these problems, a socio-technical approach is necessary, and it implies smart automation and human control accompanied by the ethical governance system. Integrating explainability, fairness, and privacy-by-design will make cognitive systems cognitively transparent, fair, and aligned with societal values. In addition, the use of human-in-the-loop systems allows to keep people both responsible and free, especially in those areas where an AI mistake can potentially pose economic or ethical damages. New technology like blockchain to verify the identity of security devices, quantum-proof encryptions, and generative AI to simulate threats, are potential solutions, but introduce new dual-use, governance, and ethical issues that should be addressed through collaboration and standard set harmonization on an international scale [21]. Moreover, the surveys that have been conducted globally show that people are only apprehensive in their trust in AI applications- although they agree with technical prowess in their usage, they require openness, agency, and ethicalness in accordance with human goals. According to the WEF, cyber resilience is a step-by-step journey and as such is not a destination, one which is underpinned by trust, transparency, and transformation. In other words, the vision of cognitive security that is needed in the Industry 4.0 has to be visionary in the terms other than technical excellence: based in human values, active governance, and just systems. And only then will our intelligent systems be able to protect not only operational processes of the industry, but also people, principles, and future that they eventually serve.

The need of human-in-the-loop frameworks can hardly be exaggerated, as they maintain the aspects of agency and accountability when mistakes have a risk of creating operational, economic, or ethical damages and disasters. Furthermore, the secure-by-design architectures should become a standard so that IIoT systems were authoritatively surviving, interoperating, and demonstrating flexibility. This encompasses the thoughtful incorporation of novel technologies, including blockchain in the field of verifiable device identity, quantum-resilient encryption and generative AI in simulating threats. However, the use of such tools poses multifaceted challenges concerning dual use, governance, and ethical risk- which requires international collaboration, harmonization of the standards and multistakeholder enforcement. The path toward obtaining Industry 4.0 is not a fixed target, and it is a moving process that requires collective responsibility and inter-sectoral effort. It is a common struggle as the World Economic Forum adequately observes, that cyber resistance is not a destination, but rather a collective journey, which needs trust, transparency and transformation. Cognitive security can only be reached with a vision that extends beyond technical aspiration it should be based on human ideals, active prevention and system equality. Only in this way, we can be sure that the intelligent systems we create would not only safeguard industry, but also the people, principles, and future they are steward to.

REFERENCES

1. Yin, C., Zhou, H., Tang, J., & Liu, L. (2023). An intelligent detection approach for industrial cyber threats using multi-layer CNN-LSTM. *IEEE Access, 11*, 56632–56645. https://doi.org/10.1109/ACCESS.2023.3271411
2. Floridi, L., Cowls, J., Beltrametti, M., Chatila, R., Chazerand, P., Dignum, V., ... & Vayena, E. (2018). AI4People—An ethical framework for a good AI society: Opportunities, risks, principles, and recommendations. *Minds and Machines, 28*(4), 689–707. https://doi.org/10.1007/s11023-018-9482-5
3. UNESCO. (2021). *Recommendation on the Ethics of Artificial Intelligence*. https://unesdoc.unesco.org/ark:/48223/pf0000380455
4. European Commission. (2024). *Proposal for a regulation laying down harmonised rules on artificial intelligence (Artificial Intelligence Act)*. https://digital-strategy.ec.europa.eu

5. Shakya, S. R., & Jha, S. (2022). Challenges in Industrial Internet of Things (IIoT). In S. Jha, U. Tariq, G. P. Joshi, & V. K. Solanki (Eds.), *Industrial Internet of Things: Technologies, Design, and Applications* (pp. 21–40). CRC Press. https://doi.org/10.1201/9781003102267-2
6. Yadav, R., & Sharma, P. (2024). Deep learning-based security architecture for IIoT: A hybrid model. *Cyber Physical Systems Review, 12*(1), 41–58.
7. Li, T., Sahu, A. K., Talwalkar, A., & Smith, V. (2020). Federated learning: Challenges, methods, and future directions. *IEEE Signal Processing Magazine, 37*(3), 50–60. https://doi.org/10.1109/MSP.2020.2975749
8. Gunning, D., Stefik, M., Choi, J., Miller, T., Stumpf, S., & Yang, G.-Z. (2019). XAI—Explainable artificial intelligence. *Science Robotics, 4*(37), eaay7120. https://doi.org/10.1126/scirobotics.aay7120
9. Doshi-Velez, F., & Kim, B. (2017). *Towards a rigorous science of interpretable machine learning*. arXiv. https://arxiv.org/abs/1702.08608
10. Ribeiro, M. T., Singh, S., & Guestrin, C. (2022). Why should I trust you? Explaining the predictions of any classifier. Proceedings of the 22nd ACM SIGKDD, 1135–1144. https://doi.org/10.1145/2939672.2939778
11. Jha, S., Nkenyereye, L., Joshi, G. P., & Yang, E. (2020). Mitigating and monitoring smart city using Internet of Things. *Computers, Materials & Continua, 65*(2), 1059–1079. www.techscience.com/cmc/v65n2/39862
12. Mittelstadt, B. D., Allo, P., Taddeo, M., Wachter, S., & Floridi, L. (2016). The ethics of algorithms: Mapping the debate. *Big Data & Society, 3*(2), 1–21. https://doi.org/10.1177/2053951716679679
13. Ajunwa, I., Crawford, K., & Schultz, J. (2017). Limitless worker surveillance. *California Law Review, 105*(3), 735–776. https://doi.org/10.15779/Z38BR8MF94
14. Jha, S., Tariq, U., Joshi, G. P., & Solanki, V. K. (Eds.). (2022). *Industrial Internet of Things: Technologies, Design, and Applications*. CRC Press.
15. Brundage, M., Avin, S., Wang, J., Belfield, H., Krueger, G., Hadfield, G., ... & Dafoe, A. (2023). *Trustworthy AI: A Review of Challenges and Research Directions*. Partnership on AI. https://partnershiponai.org
16. Amodei, D., Olah, C., Steinhardt, J., Christiano, P., Schulman, J., & Mané, D. (2016). *Concrete problems in AI safety*. arXiv. https://arxiv.org/abs/1606.06565
17. Kairouz, P., McMahan, H. B., Brendan, A., Aurélien, B., Mehdi Bennis, Arjun, N. B. ... & Sen, Z. (2021). Advances and open problems in federated learning. *Foundations and Trends® in Machine Learning, 14*(1–2), 1–210. https://doi.org/10.1561/2200000083
18. World Economic Forum. (2023). *Global Cybersecurity Outlook 2023*. www.weforum.org/reports/global-cybersecurity-outlook-2023
19. Jha, S., Ahmad, S., Abdeljaber, H. A. M., Othman, N. A., Nazeer, J., & Yadav, K. (2025). Enabling resilient wireless networks: OFDMA-based algorithm for enhanced survivability and privacy in 6G IoT environments. *IEEE, 70*(1), 3810–3819. https://ieeexplore.ieee.org/abstract/document/10445489
20. Kroll, J. A., Huey, J., Barocas, S., Felten, E. W., Reidenberg, J. R., Robinson, D. G., & Yu, H. (2016). Accountable algorithms. *University of Pennsylvania Law Review, 165*, 633–705.
21. Fernández-Caramés, T. M., & Fraga-Lamas, P. (2019). A review on the application of blockchain to the next generation of cybersecure Industry 4.0 smart factories. *IEEE Access, 7*, 45201–45218. https://doi.org/10.1109/ACCESS.2019.2908780

4 Behavioral Analytics for Threat Detection

T. Kirthiga Devi and G. Maragatham

4.1 INTRODUCTION

4.1.1 OVERVIEW OF BEHAVIORAL ANALYTICS THREAT DETECTION

The behavioral analytics approach to threat detection is a data-driven paradigm that capitalizes on statistical profiling, machine learning (ML), and deep learning algorithms to identify the user and entity behavior patterns that indicate potential threats. Unusual behavior patterns that pose a high-risk security threat, such as insider attacks, compromised accounts, or information or data breaches. This approach derived from static models, rule-based detection models to dynamic behavioral monitoring and baselining, wherein systems identify the profiles from vast volumes of data, logs, endpoints, network traffic, and application usage [1–4].

The core of behavioral analytics (Figure 4.1) focuses on User and Entity Behavior Analytics (UEBA), which utilizes different models of unsupervised learning, clustering techniques, and pattern matching and recognition to flag the alert of anomalies based on established baselines. In particular, numerous reports on privilege escalations, failed login attempts, or abnormal data transfer in the network or through email are flagged as indicators of compromise (IoCs). These anomalies are prioritized based on risk scoring models by analyzing data correlated with contextual metadata like who the user is, what device they're using or assigned to, and what time the activity occurred to reduce false positives [5–8].

Seamlessly integrating security information & event management (SIEM) systems, graph-based analysis, and behavioral biometrics with behavioral analytics for collecting high volumes of security data across organizations to enhance visibility into multi-vector threats such as insider attacks, lateral movement, and credential misuse. As it evolves in adapting over time, the model provides a scalable, adaptive framework for detecting persistent threats in real-time, significantly to improve the key factors Mean Time to Detect (MTTD), Mean Time to Respond (MTTR), Mean Time to Contain (MTTC), Mean Time to Remediate (MTTR), and Mean Time to Recover (MTTR) metrics.

DOI: 10.1201/9781003609773-4

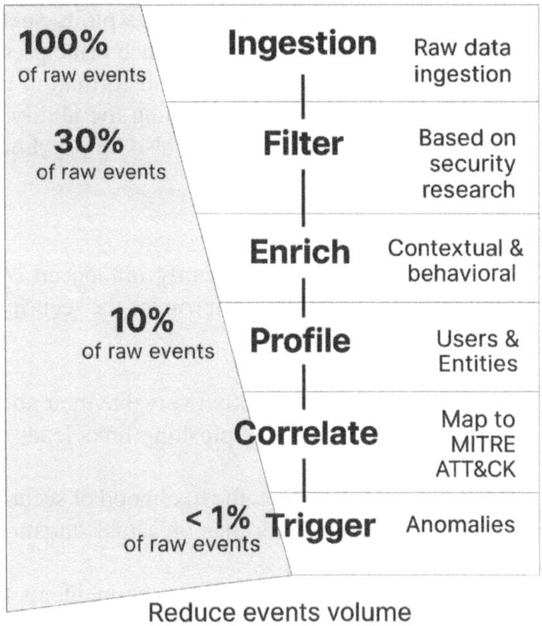

FIGURE 4.1 Detail analysis of threat detection.

4.1.2 Importance of Understanding User and System Behaviors

It's fundamental to understand user and system behaviors, which are flagged as critical in threat detection, as many threats are classified based on intentional or unintentional patterns. User behaviors can inadvertently or deliberately introduce vulnerabilities, ranging from accidental data leaks to organized and planned insider threats [9–12]. The system behavior is categorized as misuse of app data, unlimited downloads, data access patterns, modify system activities, accessing data on nonworking hours, login attempts, increased system/network traffic, violating policies or network traffic anomalies, often signals potential cybersecurity incidents [13]. The organization's ability to predict, detect, prevent, and immediately respond to security threats is restricted to satisfying guidelines and compliance in designing the model to minimize risk and protect critical assets [14–16].

4.2 FOUNDATIONS OF BEHAVIORAL ANALYTICS

4.2.1 Behavioral Science and Its Relevance to Security

Behavioral science gives detailed insight into how individuals make decisions and interact within social and cognitive dynamic contexts [17–20]. It provides a foundational layer for behavioral analytics, enabling strategies to discover the mental heuristics, decision-making flaws, psychological tendencies, malicious/benign intent, and social dynamics as security systems anticipate in identifying the threats, such as insider attacks, phishing, and social engineering exploits such as manipulation tactics and threat vectors [21]. When behavioral science principles of cognitive bias, social authority, heuristics, risk perception, and habit formation are embedded into analytical models, organizations narrow in on firming the ability to flag suspicious actions not just based on technical anomalies, but also on psychologically informed deviations [22, 23].

Cognitive Biases and its Risks

Decision-making of every individual is prevalently influenced by cognitive biases which results in risky behavior pattern that compromise the security. The key relevant patterns are [24]:

- **Overconfidence Bias:** Users frequently overvalue their ability to identify the threat such as employee clicking the phishing links leads to losing sensitive data [25].
- **Availability Heuristic:** Users assess the likelihood of security incidents based on how they can recall events rather than on actual statistical data, which can misrepresent risk perception [26].
- **Confirmation Bias:** Individuals investigate the incidents to gather information that supports their theory, making them susceptible to misinformation or targeted manipulation [27].

Social Engineering and Psychological Manipulation

Cybercriminals don't require bypassing and exploiting technical vulnerabilities to access sensitive data but can manipulate human behavior to gain unauthorized access to systems or sensitive information by taking advantage of trust, fear, and curiosity. The social engineering strategies include

Phishing: It involves emails with malicious links intended to trick users into giving credit card details, login credentials, or banking details.

Pretexting: It often occurs when attackers pretend to be trustworthy, authorized individuals, like technical support people or managers, to manipulate users into disclosing sensitive data.

Vishing: Termed as voice phishing, it involves fraudulent calls where attackers impersonate a legitimate entity to deceive users into revealing sensitive data.

Shoulder surfing: It refers to the act of physically observing users entering credentials, passwords, or PINs in public environments.

Dumpster Diving: It involves searching through the discarded documents, hardware, and software to get valuable information. The implication to remind data protection must go beyond the digitalized documents to include physical disposal protocols.

Baiting and Quid Pro Quo: In these attacks, an offer is made in exchange for confidential information or system access to the network. The interaction is more direct; often, the perceived individual as a disgruntled employee or in need of financial stability will communicate over phone calls or in person.

4.2.2 Key Concepts in Behavioral Analytics

Behavioral analytics involves developing systems that continuously monitor behavioral patterns of user and system, that detect potential anomalies, predicts security events, and mitigates threats.

- **Behavioral Baselines**: A behavioral baseline is a dynamic process that represents in identifying who and what to monitor: entities and behavior types. The entities are broadly classified into human-centric, system or technical, adversarial, organizational, abstract or logical in threat detection that can be modeled accurately. The major step involves collecting a large variety of data to denote behavior types with its vectors such as logins, file access, data transfers, application usage, incoming and outgoing email etc. The baseline is created based on the historical pattern collected over a period of time to establish the contextual thresholds that evolve with changing patterns.
- **Identifying Normal versus Abnormal Behavior:** Collected data is modeled to construct behavioral profiles using algorithms that recognize deviations from normal patterns. Live data is then compared to these baseline profiles to identify anomalies based on the magnitude and context. Context models help differentiate benign deviations from suspicious actions.
- **User Risk Scoring:** To quantify risk scoring, each indicator is mapped to risk vectors aligned with policy and threat models. Risk values are then assigned using expert- or data-driven models by categorizing users' behaviors into adaptive thresholds such as low, medium, or high risk. The scoring system gets updated on threat intelligence, incident outcomes, and audit results. This approach helps reduce false negatives and false positives over time.
- **Behavior-based Authentication:** To authenticate users behavior profiles, signals are identified through keystroke dynamics, mouse movements, and browsing patterns and get modeled to reduce noise and session segmentation. Training individual patterns helps to get the behavioral signature of each user to ensure reliability. User behavior during sessions is constantly verified with existing profiles based on thresholds in such a way that behavior may evolve.
- **Anomaly Detection**: Anomaly detection helps recognize patterns in data attributes, behavior patterns, and events that deviate from baselines. These anomalies indicate key factors concerning security policy violations, compromised accounts, security breaches, or any live incident.
- **Risk Scoring**: Risk scoring uses quantitative methods to assess the impact and severity of anomalies and security incidents based on contextual factors such as historical behavior, external access level, and privilege indicators [28]. The model helps prioritize alerts; the effectiveness of incident response operations supports threat mitigation based on the risk assessment. These components work together to provide a comprehensive analytical framework for organizations to detect, prioritize, and respond to threats.

4.2.3 The Role of Data in Understanding Behaviors

The term 'data' is the core foundation of behavioral analytics that enables organizations to observe, collect, integrate, interpret, and respond to every individual user's actions. One of the major challenges exists in integrating data across various sources due to heterogeneous data structure, time synchronization issues, semantic inconsistency, and regulatory constraints. The crucial factor resolving issues lies in the preprocessing pipeline and feature extraction. The feature engineering types are broadly inclined toward temporal, frequency, peer comparison, location, and cumulative risk.

Types of Data Used in Behavioral Analytics

System Logs: These logs capture events and track user behavior patterns through key fields: user_id, event_type, timestamp, hostname, location, resource_accessed.

Network Traffic: These logs track the inbound and outbound raw packet data and flow-level data from internal to external and vice versa. Then raw traffic is structured to correlate live patterns. Traffic often signals the pattern that helps to detect unauthorized access or policy violations. The network key fields are src_ip, dst_ip, port, protocol, bytes_sent, and connection_duration.

Biometric Data: It establishes user identity through physiological and behavioral traits that strengthen the authentication mechanism. The data key fields are typing speed, facial expression, voice patterns, rhythm, and deviations.

Application Usage Data: It records how users interact with the software platform through features like command sequence, role-to-activity, and module frequency.

Data Collection and Privacy Considerations

The organization collects sensitive data to study user behavior and identify potential risks. The initial step to define the objective on which behavior needs to be monitored improves detections. The ethical boundaries are then implemented, adhering to regulations like GDPR and HIPAA, by formalizing policies and enhancing a consent mechanism.

Category	Details & Examples	Regulations & Policies	Best Practices
Data Collected	User behavior, network logs, app usage, system events, HR records, physical logs	GDPR, CCPA, HIPAA, SOX, FISMA, FedRAMP, NIST	Data minimization, anonymization, consent
Privacy Risks	Excessive monitoring, misuse, discrimination, transparency concerns	GDPR, CCPA, HIPAA, SOX, NIST	Clear policies, transparency, audits
Stakeholders	Threat managers, DPOs, security analysts, legal teams, HR representatives	GDPR, SOX, HIPAA, FISMA, NIST	Role clarity, privacy training, controlled access

Category	Details & Examples	Regulations & Policies	Best Practices
Why Important	Mitigate insider risks, legal compliance, employee trust, operational continuity	GDPR, CCPA, HIPAA, SOX, NIST	Ethical data use, compliance reviews
Tech & Tools	UEBA, DLP, SIEM integration, AI-based detection	GDPR, HIPAA, SOX, FISMA, NIST	Privacy controls, encryption, audit trails
Framework Steps	Purpose definition, transparency, proportionality, secure storage, controlled access, retention policies	GDPR, CCPA, HIPAA, SOX, NIST	Privacy by design, clear consent, breach plans
Future Trends	AI privacy controls, evolving regulations, employee engagement	GDPR, CCPA, SOX, HIPAA, NIST	Continuous improvement, adaptive privacy strategies

Analyzing Data for Security Insights

Advanced data analytics methodology helps enhance detection strategy in identifying the threats and corresponding behavioral indicators of live data collection through behavioral clustering, sentimental analysis, time-series analysis.

- **Behavioral Clustering:** This method uses data to group users based on similar behavior patterns and actions within the organization, to detect malicious or negligent insiders and it becomes a lot easier to identify risks and reduce false alarms. For example, in healthcare, organizations use this approach to track when medical staffs access patient records across various departments who has different privilege access during work hours.
- **Sentiment Analysis:** In the insider threat context, sentiment analysis uses natural language processing to interpret emotional cues like stress, anger, resentment, and dissatisfaction from users' activity—such as emails, internal chats, performance reviews, and social media activities. These are evaluated through sentiment scoring, which helps trigger alerts if negative emotions escalate or correlate.

4.3 TYPES OF THREATS AND BEHAVIORAL INDICATORS

Security threats have evolved significantly over time, making it harder to dissect the evolution of threats and their behavioral indicators. In the broader landscape, it's essential for an organization to observe emotional cues that may help indicate malicious intent, especially when threat patterns are subtle or not. This section will explore cybersecurity, physical, and hybrid threats associated with indicators to examine their nature, impact, and their role in mitigating risk.

Threat Type	Sub-Category	Detailed Behavioral Indicators	Behavioral Analytics Response
Cybersecurity Threats	**Insider Threats**	- Sudden access to confidential or unrelated data- Access outside working hours or from unusual locations- Unexplained use of admin privileges	Behavioral baselining detects deviations from normal access behavior. UEBA tools flag high-risk profiles and correlate actions with identity, time, and device history to detect insider threats.
	Phishing	- Increased clicking on links in emails- Responding to urgent or suspicious emails- Input of credentials into untrusted sites	Advanced analytics monitor communication patterns and flag behavior triggered by social engineering techniques. Email behavior modeling detects anomalies in sender-recipient relationships and link-clicking patterns.
	Malware	- High volume of outbound data transfers- Unusual processes running on a system- Rapid system file access or encryption (e.g., ransomware behavior)	Endpoint and network behavioral analytics detect abnormal system activities. Anomaly detection models identify deviation in file system use, network traffic volume, and unexpected application behavior.
Physical Security Threats	**Suspicious Activities**	- Presence in restricted areas without clearance- Extended loitering near secure locations- Unauthorized entry or tailgating	Physical surveillance data (CCTV and access card logs) is analyzed to detect time-location anomalies. Behavioral profiling detects unusual movement or unauthorized access patterns.
	Workplace Violence	- Escalating conflicts with coworkers- Aggressive gestures or verbal threats- Sudden behavioral change in demeanor	Behavioral observation systems and HR-flagged behavioral trends are analyzed to predict escalation. Real-time monitoring alerts security staff to intervene and reduce harm.
Hybrid Threats	**Converged Physical and Cyber Behaviors**	- Unauthorized physical entry combined with system access- Use of stolen credentials following physical breach- Simultaneous badge-in with VPN login anomalies	Cross-domain analytics integrates physical access logs with IT activity (e.g., badge swipe + login). Multi-modal correlation identifies threats that span both cyber and physical dimensions, enabling coordinated mitigation across departments.

4.3.1 Cybersecurity Threats: Insider Threats, Phishing, and Malware

Cybersecurity threats are the most prevalent and evolving risks in today's modern era. **Insider threats** may involve individuals in an organization, such as employees or contractors, who misuse their privileges to destroy the reputation of the organization or compromise the integrity. The person may act with malicious intent or unintentionally leads to security breaches due to negligence or mistakes. Indicators of insider threats include various factors like unauthorized access to sensitive files, excessive downloading during nonworking hours, or sudden changes in behavior due to emotional conflict with coworkers or with the organization.

Malware is malicious software that infects enterprise networks to steal data or disrupt operations. Behavioral indicators of a malware attack are unusual system slowdowns, unauthorized software installations, and recurrent failures.

4.3.2 Physical Security Threats

Physical security threats present dangers to organizations and individuals, leading to long-term disruptions. Suspicious activities include behavior patterns that deviate from normal, such as unauthorized access attempts or individuals who avoid security checkpoints. The ongoing research helps recognize those behaviors early, which can prevent security breaches and potential risks due to attacks.

Workplace violence is a critical threat that includes an act of aggression, physical assaults within an organization setting. The key behavioral indicators of workplace violence include frustration, verbal threats, withdrawal from interactions, or aggressive behavior toward coworkers or senior employees. To address this, proactive measures are adopted by organizations to mitigate workplace risk in organizations.

4.3.3 Hybrid Threats: Convergence of Physical and Cyber Behaviors

A hybrid security threat involves the integration of multiple attack vectors, like cyber, physical, and psychological, that leads to strategic disruption. The characteristics of a threat are classified as multi-vector, stealth, persistence, and motives. To address this, a cross-domain structured intelligence approach is applied in creating a baseline, threat modeling, risk scoring, and integration strategies to enhance the preparation of forensic readiness.

The behavioral indicators of hybrid threats (Figure 4.2) includes unusual online activity linked to physical actions, for instance, an employee accessing restricted areas after interacting with external threat actors via email. The integration of cybersecurity and physical security measures helps organizations to detect and prevent such hybrid threats effectively.

4.4 THE BEHAVIORAL ANALYTICS PROCESS

Behavioral analytics is a systematic approach that helps to monitor, detect and design responses to severity-level classified threats based on pattern behavior. It involves

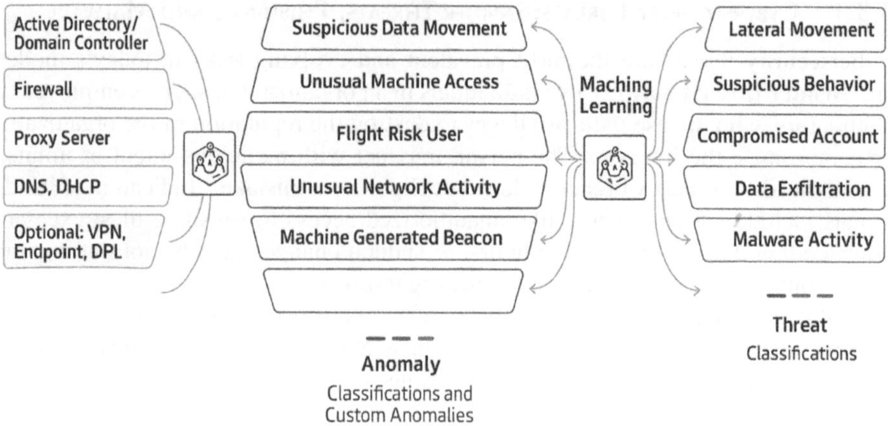

FIGURE 4.2 Behavior analytics.

analyzing data from multiple sources to identify deviation in collected patterns to indicate potential risks. It's important in identifying insider threats, cyberattacks, and physical security risks which may escalate into major incidents.

4.4.1 Data Collection and Aggregation

The foremost step in behavioral analytics is data collection and aggregation. It involves gathering data from multiple sources, including digital logs, physical surveillance systems, and user activity records. The collected data is then aggregated into a centralized system for analysis. Effective data collection allows security teams to establish a comprehensive understanding of normal and abnormal behaviors, making it easier to detect potential threats.

To detect threats, organizations must collect and aggregate various multimodal data from multiple sources, such as

- Access logs
- Device usage patterns
- Physical entry records

4.4.1.2 Sources of Data: Logs, Sensors, Social Media, and More

To conduct behavioral analytics effectively, organizations must collect data from diverse sources. Logs from computer systems offer details about user activity, such as login attempts, file access, and software usage. Sensors, such as motion detectors and biometric scanners, capture physical movements and access patterns. Social media platforms can also be valuable sources, as they may reveal threatening posts, discussions about insider threats, or indicators of upcoming attacks. By leveraging these varied sources of data, security teams can develop a more holistic understanding of potential threats.

4.4.2 Behavioral Baselines: Defining "Normal"

To detect anomalies effectively, it is essential to establish behavioral baselines, which define what constitutes "normal" activity within a given environment. A baseline is created to analyze historical data by identifying known patterns in user behavior, network activity, and physical movements. Once a baseline is established, deviations from normal behavior can be flagged as potential security threats.

For example, if an employee typically accesses a company's systems during working hours but suddenly starts logging in at unusual times or from different locations, this deviation from their normal baseline may indicate a security risk. Similarly, if a visitor repeatedly lingers near restricted areas, their behavior may be flagged as suspicious. Creating a baseline for normal behavior involves analyzing historical data to determine patterns typical of individuals or systems.

4.4.3 Identifying Anomalies: Detecting the Deviations

The final stage in the behavioral analytics process is identifying anomalies, which involves detecting deviations on the existing established baseline that certainly indicate potential threats. Anomalies areclassify into categories, such as statistical anomalies (unusual spikes in activity), contextual anomalies where behavior that is unusual in a specific situation, and unified anomalies referred to as a group of behaviors suggests a security risk.

For example, an employee suddenly transferring large amounts of sensitive data outside the organization's network is a clear anomaly that requires investigation. Similarly, an individual repeatedly bypassing security protocols in a physical location could indicate malicious intent. They are often used to improve anomaly detection by analyzing large amounts of data and identifying complex patterns that analysts may overlook. UsingML, statistical models, or predefined rules, organizations can detect when behavior deviates from the baseline, potentially signaling a threat.

Data are the foundational input for behavioral analytics, enabling the construction of detailed and dynamic behavioral profiles.

- **Network and system logs**
- **Physical surveillance data**
- **Communications** over the network
- **User activity monitoring**

By amalgamating multimodal data streams, organizations detect anomalies and help to collect high-quality metadata [29], which enables them to improve analytics from reactive to proactive threat mitigation, offering deep vertical behavioral insights.

4.5 INTRODUCTION TO BEHAVIORAL ANALYTICS FOR THREAT DETECTION

In the modern digital era, there is a rapid evolution of technology and cyber threats, so the need for robust, adaptive, and intelligent systems becomes mandatory. The methodology of analyzing user behavior for threat detection is gaining momentum, leveraging techniques and computational intelligence to derive behavioral insights in real time. The main focus on patterns, anomalies, and deviations in behavior is used to detect known and unknown threats in real time to provide proactive defense.

4.5.1 The Evolution of Threat Detection

The traditional methodology uses a signature-based system and static rule set to detect threats. The known threats are identified with limited attack vectors. The rise of zero-day attacks, insider threats, and social engineering tactics has exposed the inability and inadequacies of conventional security frameworks. Behavioral analytics offers a transformative shift by moving from static, rule-based detection to dynamic, context-aware approaches that analyze user, entity, and system behaviors in real time.

The studies conducted by Liao et al. (2020) and Sharma et al. (2021) have highlighted the value of behavioral analytics in uncovering hidden patterns that traditional methods might ignore. Combining different types of data—like network traffic, user activity logs, and endpoint telemetry—with behavioral analytics systems can create a complete picture of normal behavior, helping to spot small changes that might suggest harmful actions.

4.5.2 The Role of ML and AI

They have become cornerstones of modern behavioral analytics. These technologies allow for the processing and analysis of vast datasets, extracting meaningful insights and identifying anomalies with unprecedented speed and accuracy. Techniques such as unsupervised learning, supervised classification, and deep learning models have been instrumental in advancing the field.

For example, unsupervised learning algorithms like k-means clustering and isolation forests have proven effective in detecting outliers that signify abnormal behavior. Meanwhile, supervised learning models get trained on labeled datasets, to classify each activity as a benign or malicious pattern with high precision values. Deep learning techniques, integrated with recurrent neural networks and convolutional neural networks, focus on strengthening detection & helping to evolve the threats by capturing temporal and spatial dependencies in behavioral data.

4.5.3 Key Concepts of Behavioral Analytics in Threat Detection

Behavioral analytics finds applications across various domains [30], including

1. Insider Threat Detection: Identifying anomalous activities by legitimate users that indicates malicious intent, like as data exfiltration or privilege abuse.
2. Fraud Prevention: Detecting fraudulent transactions in financial systems by analyzing patterns of user behavior, such as transaction frequency, location, and device usage.

3. Network Security: Monitors network traffic to identify usual/unusual patterns, such as data frequency spikes, unusual access times, or communication with known malicious domains.
4. Endpoint Protection: Analyzing endpoint activity for deviations that indicates malware infections or unauthorized access attempts.

4.6 KEY TECHNIQUES IN BEHAVIORAL ANALYTICS

- **Anomaly Detection**: The model is designed to identify the deviations from established behavioral baselines and predicts unusual patterns that signify malicious activity. The methods included are Z-Score, isolation forest. SVM, Encoders.
- **Pattern Recognition**: It involves mining historical data to identify the recurrent behavioral patterns. The techniques that help in detection such behaviors are credential stuffing, lateral movement, privilege escalation.
- **Cluster Analysis**: It groups users' activities based on similarity. The Algorithm includes K Means, DBSCAN, hierarchical clustering.
- **Predictive Modeling**: It uses historical data to forecast future events and mitigate the risk proactively.
- **Graph Analysis**: It models the relationship between user, systems, events. The detected patterns evaluated based on influence analysis, graph traversal algorithms, subgroup identification.

4.6.1 Machine Learning and Artificial Intelligence

They play a pivotal role in behavioral analytics by automating the detection and response. ML systems refine predictive model from historical data, while AI provides cognitive capabilities for better reasoning and decision-making.

Supervised versus Unsupervised Learning

- **Supervised Learning**: The training models use labeled datasets when input and output are known. The common algorithms are decision trees, random forests, support vector machines (SVMs). This approach is highly effective for classification tasks such as detecting malware or phishing attacks.
- **Unsupervised Learning**: This model involves unlabeled data to identify hidden patterns or structures. Various clustering techniques are used to identify anomalies without prior knowledge of potential threats.
- **Semi-Supervised Learning**: This hybrid approach combines both labeled and unlabeled data, to improvise the accuracy and to reduce the cost of manual data labeling.

Neural Networks for Pattern Recognition

Neural networks are widely used for pattern recognition in behavioral analytics. The key neural network architectures include the following

- **Convolutional Neural Networks (CNNs)** are used in image processing but have been adapted for detecting behavioral threats.

- **Recurrent Neural Networks (RNNs)** are effective for analyzing sequences, especially in detecting fraudulent transactions or recurrent procedures involved.
- **Autoencoders are used for** anomaly detection to identify the deviations from normal behavior by various compressed representations.

4.6.2 Statistical Models

Statistical techniques form the foundation of many behavioral analytics models. Common approaches include

- **Bayesian Networks**: Probabilistic models used for decision-making under uncertainty.
- **Markov Chains**: Sequence models that predict future behavior based on previous states.
- **Regression Analysis**: Used for forecasting and understanding relationships between different behavioral variables.
- **Hypothesis Testing**: Helps in determining whether observed anomalies are statistically significant or random occurrences.

4.6.3 Rule-Based Systems

Rule-based systems are predefined if-then to identify security threats. They are often integrated into Security Information and Event Management (SIEM) systems and include:

- **Expert Systems**: Relying on domain knowledge to detect predefined attack patterns.
- **Heuristic-Based Approaches**: Defining heuristic rules that adapt to evolving threat landscapes.
- **Signature-Based Detection**: Matching user behavior against known attack signatures.

4.6.4 Behavioral Graph Analysis

Graph analytics is used to model relationships between users, entities, and events. This technique is instrumental in:

- **Detecting Insider Threats**: Identifying abnormal access patterns within an organization.
- **Network Traffic Analysis**: Uncovering coordinated attacks by mapping out malicious interactions.
- **Fraud Detection**: Tracking transaction relationships to detect financial fraud.
- **Social Network Analysis**: Evaluating connections between users to uncover potential security risks.

4.7 TECHNOLOGIES AND TOOLS FOR BEHAVIORAL THREAT DETECTION

To effectively implement behavioral analytics in cybersecurity, various technologies and tools are utilized:

4.7.1 Security Information and Event Management (SIEM) Systems

SIEM tools help to analyze security logs and event data from multiple sources. The tools include:

- **Splunk**: Provides real-time monitoring, searching, and visualization.
- **IBM QRadar**: Uses AI-powered threat intelligence for security event correlation.
- **ArcSight**: Offers advanced analytics for threat detection and compliance management.

4.7.2 User and Entity Behavior Analytics (UEBA) Tools

UEBA solutions focus on detecting abnormal user and entity behavior. Leading tools include

- **Exabeam**: Uses ML for risk scoring and anomaly detection.
- **Varonis**: Monitors file access and insider threats.
- **Microsoft Defender for Identity**: Detects suspicious behavior across Active Directory environments.

Behavioral analytics for threat detection is a powerful tool, but it comes with several challenges and limitations. These include:

- **Data Quality & Availability**: The accuracy depends on quality and comprehensive datasets whereas incomplete / biased data can lead to inaccurate predictions.
- **Computational Complexity**: Analyzing large volume amounts of behavioral data in real time requires significant computational power, which can be costly and resource-intensive.
- **Interpretability Issues**: Many behavioral analytics models, particularly those powered by AI, operate as "black boxes," making it difficult for security teams to understand their decision-making process.
- **Adaptability to New Threats**: While behavioral analytics is effective in identifying known attack patterns, it may struggle to detect novel threats that deviate from existing behavioral baselines.

4.8 CHALLENGES AND LIMITATIONS

4.8.1 Data Privacy and Ethical Considerations

Behavioral analytics involves the collection and analytics of processing sensitive data. Below are the ethical and privacy concerns included:

- Informed Consent: Users being aware of guidelines and understanding how their data is collected, processed, and utilized.
- Data Minimization: Organizations foster collecting only the required data that helps to reduce the risk and prevent mishandling of data.
- Anonymization Techniques: used like differential privacy can help protect user identities while enabling effective analysis.
- Fairness and Bias Mitigation: Behavioral analytics models should be trained and evaluated to minimize bias and prevent discriminatory outcomes.

4.8.2 False Positive and False Negative

- False Positive (False Alarms): These occur when legitimate activities are flagged to be unusual suspicious activities, leading to unnecessary investigations.
- False Negative (Missed Threats): These happen when actual high-level threats go undetected and lead to data breaches, financial losses, or reputational damage.
- Strategies for Improvement: The objective to enhance data quality, refine the algorithms, and incorporate human behavior patterns can help to mitigate these issues.

4.8.3 Scaling Behavioral Analytics Systems

- Infrastructure Requirements: The system ensures storage, processing capability, and required network capacity to handle the increased data loads.
- Latency Concerns: Real-time analytics must be adaptive to new learning technologies as data grows.
- Automated Threat Response: AI-driven systems are implemented to analyze and respond to anomalies without any intervention.

4.8.4 Evolving Threat Landscapes

As cyber threats evolve, it's important for behavioral analytics to adapt, which includes

- AI-Powered Attacks: The adversaries are leveraging an AI to bypass known security mechanisms.
- Insider Threats: Employees or contractors with privileged access pose a security risk.

- Advanced Persistent Threats: Continuous monitoring of zero-day attacks is highly advisable to keep in check.

4.9 FUTURE OF BEHAVIORAL ANALYTICS FOR THREAT DETECTION

The field of behavioral analytics is going through a major shift, enabled by innovation and advancements in adaptive learning of artificial intelligence, real-time monitoring, and implementation of federated learning. These developments can improve drastically to enhance early stage threat detection and real time dynamic risk mitigation.

Category	Technical Advancement	Key Benefits / Impact
AI/ML Platforms	AutoML, federated learning, transformers	Automated behavior learning; scalable across environments
Cloud & Edge Computing	Serverless analytics (e.g., AWS Lambda), edge ML on devices	Low-latency, real-time detection; works in constrained environments
Behavioral Biometrics	Keystroke dynamics, gait, mouse movement, voice cadence	Continuous user verification; prevents impersonation
Blockchain	Immutable behavioral logs, smart contracts	Tamper-proof forensic trails and audit trust
Process Automation	Adaptive baselining, SOAR integration, behavioral simulations	Dynamic thresholds, automated response, proactive modeling
Graph-Based Models (GNN)	User-device-resource relationship graphs	Detects hidden behavioral patterns like lateral movement or collusion
Context-Aware Models	Embeds role, location, time, historical context	Reduces false positives with personalized behavior scoring
Hybrid Deep Learning	CNN + LSTM + Transformer-based architectures	Learns complex, long-term behavior sequences with precision
Self-Supervised Learning	Predicts missing patterns without labels	Cuts labeling cost; adapts to unseen behavior types
Insider Threat Sandboxes	Synthetic environments for model training	Safe testing and learning from rare threat behaviors
Anomaly Detection (Unsupervised)	Variational autoencoders, Isolation Forests, GANs	Identifies novel or zero-day behaviors without needing predefined rules
Sequential Forecasting Models	LSTM, Transformers, RL-based agents	Predicts future actions; supports policy learning and pre-breach alerts
Causal Inference Techniques	Granger causality, Bayesian networks	Reveals cause-effect links in user behavior and threat outcomes
Accuracy & Resilience	Ensemble models, online learning, drift detection	Maintains model performance over time; adapts to evolving behaviors

Category	Technical Advancement	Key Benefits / Impact
SOAR & SIEM Integration	Behavioral triggers connected to automated response workflows	Speeds up incident handling and improves decision consistency
Privacy-Preserving AI	Federated learning, differential privacy	Enables ethical modeling without violating user confidentiality
Explainability (XAI)	SHAP, LIME, LRP, interpretable models	Builds trust in automated decisions; required for compliance and forensics
Digital Twins for Users	RL-based behavior simulations mirroring user actions	Allows testing "what-if" scenarios before real-world deployment
Cognitive Behavioral Modeling	Emotion, sentiment, and stress signal integration	Detects insider threats driven by psychological distress or intent
AI SOC Co-Pilots (LLMs)	GPT-type agents assisting in behavioral analysis	Enhances analyst productivity; translates complex signals into insight

4.10 BEST PRACTICES FOR IMPLEMENTING BEHAVIORAL ANALYTICS

4.10.1 Building a Behavioral Analytics Framework

- Set measurable goals to identify the quantifiable outcomes and relevant performance benchmarks
- Construct behaviour profiles using historical data and define normative threshold based on role and context
- Ensure adapting learning pipelines are incorporated in threat intelligence and behavioral drift

4.10.2 Aligning with Organizational Goals

Behavioral analytics initiatives must integrate within the organizational architecture and goals. Whether the focus is on security, fraud prevention, customer experience enhancement, or workforce optimization, analytics must serve a strategic purpose. Leadership must perform indicators to enable the effectiveness of analytics in contributing to business objectives. Furthermore, perform scheduled module revaluations to prevent drift in model and robustness.

- Integrate security outcomes with risk management strategies
- Preserve the balance between threat mitigation and usability

4.10.3 Continuous Learning and Adaptation

To maintain effectiveness, behavioral analytics requires continuous learning to adapt to evolve attack techniques . ML models and data analysis methods should be regularly optimized to remain aligned with emerging trends and behavioral patterns. Organizations should establish real time loops where it allows insights to be monitored and adapted in real time.

- Periodically retrain models to stay aligned with dynamic threat
- Implement feedback loops to enhance threat identification accuracy

4.10.4 Collaboration Across Teams: IT, Security, and Leadership

The collaboration helps to align the socio-technical security model, where controls are strengthened by organizational processes and stakeholder communication. The responsibilities of stakeholders in insider threat management are account provisioning, log management, endpoint configuration, threat detection, SIEM correlation, incident response, risk prioritization, budget allocation, and policy formulation by respective teams — IT operations, security team (SOC, IR), and leadership (CISO, CIO, COO). The collaboration framework focuses on achieving communication channels, shared platforms, and data pipelines through cross-functional knowledge sharing.

4.11 ETHICAL IMPLICATIONS AND RESPONSIBLE USE

The implementation of behavioral analytics brings ethical considerations to the forefront—in relation to data privacy, user consent, and mitigating bias. Organizations prioritize regulating policies to establish ethical guidelines that dictate responsible data usage and ensure analytics doesn't lead to misguided or misleading user profiling. The ethical regulations prioritize openness on how behavioral data is used for maintaining user trust and meeting compliance standards for better decision-making. Ethical review boards or advisory committees oversee the changes made in policies related to analytics initiatives.

4.11.1 Balancing Security and Privacy

The biggest challenge in behavioral analytics is analysing how users behave and to balance both security and privacy aspects. The art of using behavioral data to enhance the security measures and behavioral profiling helps to address privacy concerns. Organizations must prioritize privacy approaches, implementing measures in addressing data anonymization, minimal data, retention, and secure storage. Compliance inline with dynamic data, such as GDPR and CCPA, should be considered for user right.

- Implement privacy-by-design principles to safeguard user data while maintaining security.
- Avoid intrusive surveillance that could violate user rights.

4.11.2 Transparency in Data Use

Transparency is a fundamental principle in the ethical use of behavioral analytics. Organizations should communicate how behavioral data is collected, processed as per the regulation by providing clear privacy policies . This mechanism ensures the individuals understand & have control over their data. Furthermore, offering opt-in and opt-out features strengthen the trust with stakeholders.

- Clearly communicate how behavioral data is used and protected.
- Provide users with options to control their data.

4.11.3 Legal and Regulatory Considerations

Behavioral analytics must adhere to legal and regulatory frameworks that govern data privacy and security. Regulations of GDPR, CCPA, and HIPAA set guidelines for data handling and user rights. Organizations should conduct regular legal audits and engage compliance officers to ensure adherence to relevant laws. Proactively monitoring regulatory developments can help businesses remain compliant and avoid potential legal repercussions.

- Ensure compliance with regulations of GDPR, CCPA, and various -specific standards.
- To conduct the regular audits to maintain compliances.

Transparency is a fundamental principle in the ethical use of behavioral analytics. Organizations should communicate how behavioral data is collected, processed for providing clear privacy policies and user-friendly consent mechanisms for individuals who can understand and have control on their data.

Limitations

Every cybersecurity tool has its own strengths but The problem? Cybercriminals don't try to break through where security is strongest; instead, they find the vulnerabilities that exploit the system and quietly slip through. That's why today's cyber -attacks are highly targeted, leading to limitations in system infrastructure.

To stay ahead of evolving threats, cybersecurity professionals are constantly developing new tools. But the reality: no single tool can do everything or all . The objective is to design an integrated platform, a system where technologies work together, share insights, and provide layered protection. The combined approach offers visibility, understanding, and stronger defenses. As attackers grow smarter & creative, our defense systems should evolve in intelligent decision- making.

There are three key essentials of every organization should focus on:

1. **Quality Data**: Data is everything. Even the most advanced AI or ML algorithms are useless if they're working with inaccurate or incomplete data. Clean, relevant, and timely data is the foundation of effective early threat detection.
2. **System Integration**: The cybersecurity tools shouldn't operate in silos. Tools like SIEM, UEBA must be able to communicate seamlessly. The Integration ensures faster detection, smarter correlation of events, and a quicker responses for better system.
3. **Openness and Flexibility**: It's always challenging and systems need to be open enough to integrate new tools and technologies as they emerge.

Spotting insider threats like employees accessing data they shouldn't, or acting outside their normal behavior. The biggest weakness is its difficulty in detecting threats hidden inside other forms of content, especially images. The traditional behavioral tools often miss these hidden threats so specialized tools help to dig deeper into content, not just behavior.

4.12 CONCLUSION

As cyberattacks continue to rise, organizations can't afford to rely on outdated approaches that fail to protect the network. Behavioral analytics has emerged as a game-changer which doesn't wait for threats to happen as it actively watches for warning signs by analyzing patterns or anomalies in user and system behavior before blown to full incident.

When it's integrated by AI andML, behavioral analytics becomes powerful. It can process huge volumes of data, learns to adapt over time. This makes it a crucial part of modern cybersecurity strategies.

Behavioral analytics represents a shift driven in how we think about cybersecurity from reactive to proactive, from static rules to dynamic intelligence. As innovation continues to move forward, behavioral analytics will remain one of the most important tools in protecting our digital world.

REFERENCES

1. Fuentes, J., I. Ortega-Fernandez, N. M. Villanueva, and M. Sestelo. "Cybersecurity threat detection based on a UEBA framework using Deep Autoencoders." arXiv preprint arXiv:2505.11542, 2025.
2. Ali, M. Husain, and P. Hans. "Real-time detection of insider threats using behavioral analytics and deep evidential clustering." arXiv preprint arXiv:2505.15383, 2025.
3. Tuor, A. et al. "Deep learning for unsupervised insider threat detection in structured cybersecurity data streams." arXiv preprint arXiv:1710.00811, 2017.
4. Balakrishna, Sivadi. "D-ACSM: A technique for dynamically assigning and adjusting cluster patterns for IoT data analysis." *The Journal of Supercomputing* 78, no. 10, pp. 12873–12897, 2022.
5. Jha, S., S. Routray, H. A. Abdeljaber, and S. Ahmad. "A novel approach for decision support system in cricket using machine learning." *International Journal of Computer Applications in Technology* 69, no. 1, pp. 86–92, 2022.
6. Lunardi, W. T., M. A. Lopez, and J.-P. Giacalone. "ARCADE: Adversarially regularized convolutional autoencoder for network anomaly detection." arXiv preprint arXiv:2205.01432, 2022.
7. Chen, E., S. Lockey, H. Khosravi, and N. Baghaei. "Enhancing cybersecurity through machine learning-driven anomaly detection systems." *Journal of Artificial Intelligence Research and Application* 4, no. 1, pp. 123–135, 2024.
8. Balakrishna, Sivadi, Moorthy Thirumaran, and Vijender Solanki. "Machine learning based improved Gaussian mixture model for IoT real-time data analysis." *Ingeniería Solidaria* 16, no. 1, pp. 1–30, 2020.
9. Nath, S. R. "Leveraging user and entity behavioral analysis and machine learning for log-based anomaly detection." *Global Journal of Business and Integral Security* 7, no. 1, pp. 1–10, 2023.

10. Kamal, M., and F. Bouchama. "Enhancing cyber threat detection through machine learning-based behavioral modeling of network traffic patterns." *International Journal of Business Intelligence and Big Data Analytics* 4, no. 9, pp. 1–9, 2021.
11. Balakrishna, Sivadi, and M. Thirumaran. "Programming paradigms for IoT applications: an exploratory study." In Vijender Kumar Solanki, Vicente García Díaz, and J. Paulo Davim(Eds.), *Handbook of IoT and Big Data*, pp. 23–57. CRC Press, 2019.
12. DS, Bhupal Naik, Venkatesulu Dondeti, and Sivadi Balakrishna. "Comparative analysis of machine learning-based algorithms for detection of anomalies in IIoT." *International Journal of Information Retrieval Research (IJIRR)* 12, no. 1, pp. 1–55, 2022.
13. Jha, S. "Chapter 1 – Introduction to smart soft computing in engineering: Embracing intelligence and efficiency." In *Soft Computing in Smart Manufacturing and Materials*, S. Jha, S. Datta, D. Prashar, S. R. Shakya, and V. E. Balas, Eds. Elsevier, 2025, pp. 1–10. doi: 10.1016/B978-0-443-29927-8.00001-1
14. Balakrishna, Sivadi, Vijender Kumar Solanki, Raghvendra Kumar, and M. Thirumaran. "Survey on machine learning-based clustering algorithms for IoT data cluster analysis." In *Intelligent Computing in Engineering: Select Proceedings of RICE 2019*, pp. 1195–1204. Springer Singapore, 2020.
15. Balakrishna, Sivadi, M. Thirumaran, R. Padmanaban, and Vijender Kumar Solanki. "An efficient incremental clustering based improved K-Medoids for IoT multivariate data cluster analysis." *Peer-to-Peer Networking and Applications* 13, no. 4, pp.1152–1175, 2020.
16. Jha, S. et al. "Enabling resilient wireless networks: OFDMA-based algorithm for enhanced survivability and privacy in 6G IoT environments." *IEEE Transactions on Consumer Electronics* 70, no. 1, pp. 3810–3819, 2024. doi: 10.1109/TCE.2024.3370414
17. Fahnun, U. et al. "Utilizing machine learning for anomaly detection in cybersecurity systems." *ENDLESS: International Journal of Future Studies* 7, no. 2, pp. 45–56, 2024.
18. Jha, S., N. Jha, D. Prashar, S. Ahmad, B. Alouffi, and A. Alharbi. "Integrated IoT-based secure and efficient key management framework using hashgraphs for autonomous vehicles to ensure road safety." *Sensors* 22, p. 2529. www.mdpi.com/1424-8220/22/7/2529?utm_source=chatgpt.com
19. Jha, S., D. Prashar, H. V. Long, and D. Taniar. "Recurrent neural network for detecting malware." *Computers & Security* 99, p. 102037, Dec. 2020. doi: 10.1016/j.cose.2020.102037
20. Lv, Z., D. Chen, R. Lou, and A. Alazab. "Artificial intelligence for securing industrial-based cyber–physical systems." *Future Generation Computer System* 117, pp. 291–298, 2021.
21. Kuntla, S., X. Tian, and Z. Li. "Security and privacy in machine learning: A survey." *Issues Information System* 22, no. 3, pp. 224–240, 2021.
22. Balakrishna, Sivadi, Yerrakula Gopi, and Vijender Kumar Solanki. "Comparative analysis on deep neural network models for detection of cyberbullying on Social Media." *Ingeniería Solidaria* 18, no. 1 pp.1–33, 2022.
23. Alduailij, M. et al. "Machine-learning-based DDoS attack detection using mutual information and random forest feature importance method." *Symmetry* 14, no. 6, p. 1095, 2022.
24. Agrawal, P. et al. "Metaheuristic algorithms on feature selection: A survey of one decade of research (2009–2019)." *IEEE Access* 9, pp. 26766–26791, 2021.

25. Luo, Y. et al. "Deep learning-based anomaly detection in cyber-physical systems." *ACM Computing Surveys* 54, no. 5, pp. 1–36, 2021.
26. Mendonça, R. V. et al. "A lightweight intelligent intrusion detection system for Industrial Internet of Things using deep learning algorithms." *Expert System* 39, no. 3, p. e12917, 2021.
27. Bhandari, H. C., Y. R. Pandeya, K. Jha, S. Jha, and S. Ahmad. "Exploring non-Euclidean approaches: A comprehensive survey on Graph-Based techniques for EEG signal analysis." *Journal of Advances in Information Technology* 15, no. 10, pp. 1089–1105, 2024.
28. Sánchez-Zamora, X., A. Larriva-Novo, and V. A. Villagrá. "Design and evaluation of unsupervised machine learning models for anomaly detection in streaming cybersecurity logs." *Mathematics* 10, no. 21, p. 4043, 2022.
29. Jadidi, Z. et al. "Security of machine learning-based anomaly detection in cyber physical systems." arXiv preprint arXiv:2206.05678, 2022.
30. Jha, Sudan, Sidheswar Routray, and Sultan Ahmad. "An expert system-based IoT system for minimisation of air pollution in developing countries." *International Journal of Computer Applications in Technology* 68, no. 3, pp. 277–285, 2022.

5 An Optimized ML Framework for Anomaly Detection in Edge-Enabled IIoT Networks

Biswajit Tripathy, Sujit Bebortta, Subhranshu Sekhar Tripathy, Niva Tripathy, and Subhendu Kumar Pani

5.1 INTRODUCTION

With fostering smooth connectivity and intelligent automation, the Internet of Things' (IoT) and Industrial Internet of Things' (IIoT) explosive growth has revolutionized a number of industries. However, there are now serious security issues as a result of this expansion. In 2020, 33% of all affected devices were IoT devices, according to a Nokia research, highlighting the systems' susceptibility [1]. The IIoT combined with Cyber-Physical Systems (CPS) is significantly shaped by Industry 4.0, which is defined by the nexus of operational and information technologies [2]. These technologies serve as the foundation for smart city healthcare facilities, energy grids, smart manufacturing, and intelligent transportation systems.

The implementation of Machine learning (ML)-based intrusion detection systems (IDS) proves to be an effective security tool according to [3]. When using traditional IDS signatures serve as the detection methodology but such systems can detect only known threats. The pattern detection capabilities of ML-based IDS allow them to identify both known and new security threats that appear in the data [4]. The detection capability of IoT and IIoT situations depends heavily on this essential function because their security conditions change constantly [5]. The analysis speed of ML models coupled with their capacity to process large scales of data allows them to develop new defenses when they encounter fresh data inputs. Instant real-time security evaluation is important to maintain safety in shifting security environments. ML-based IDS accomplish enhanced security by identifying legitimate activities correctly which lowers false-positive alerts [6] to free up security experts for investigating real threats. Scalability exists because such systems utilize cloud-computing platforms alongside collaborative-processing frameworks for handling rising IoT and IIoT device numbers.

Training ML systems requires sufficient data quantity as well as high-quality input data [7]. Real-world data collection challenges IIoT operations because of its diverse and dynamic industrial-process characteristics [8]. To detect anomalies in

DOI: 10.1201/9781003609773-5

IIoT networks, sophisticated feature-extraction techniques must be used because the anomalies appear as subtle reflections of normal behavior [9]. The need to deploy IDS onto IIoT devices with limited resources becomes more difficult due to restrictions on both energy use and computational complexity of ML models [10].

This research puts forward a new technique to detect anomalies in IIoT networks which combines ML algorithms with advanced border-security methods. When running this system across edge nodes it processes real-time data quickly while maintaining operational efficiency and fast responsiveness for industrial procedures. The identified dataset from the IIoT is freely accessible for conducting empirical tests on the proposed methodology. The ML-model performance reaches new heights with Optuna optimization framework because its advanced sampling techniques and pruning methods discover optimal solutions from large hyperparameter spaces. The RF model benefits from this tuning procedure which leads to enhanced edge node anomaly detection compared to different models. An improvement of 0.49% accuracy occurred after running 20 optimization trials resulting in an overall increase from 92.18% to 92.67%. The combined use of optimized ML models through edge-computing platforms leads to enhanced security performance of the IDS while providing defense against multiple types of security threats. The system demonstrates practical use in resource-limited environments which extends its functionality for performing real-time security operations within IIoT ecosystems.

5.2 RELATED WORK

Research into IoT security continues at a high-volume rate focusing on supply chain management manufacturing and logistics [11]. Many scholars have taken notice of this focus through research on possible solutions to enhance IIoT-device privacy in industrial environments by developing ML-based IDSs through different datasets [12] (see Table 5.1). Several recent datasets serve effectively both for training and evaluation of IDS models because they replicate actual network attack signatures [13]. Reliable and efficient IDS become challenging to create because these datasets display substantial class imbalances. The elimination of such problems requires three standard preprocessing procedures which include feature normalization as well as dimensionality reduction together with data balance methods [14]. The detection performance improvement in combination with maintaining accurate detection rates for imbalanced classes has led researchers to explore different ML and deep learning (DL) algorithms and feature selection methods [15].

Gad et al. [16] established a machine-learning-based detection system that serves IoT safety detection and mitigation needs. The research team utilized TON_IoT dataset as their testing and training platform because it contained information from edge, fog, and cloud layers. Researchers applied DT alongside RF and XGBoost and KNN as ML approaches for their study. The class imbalance problem was addressed using the SMOTE approach while chi-square statistics selected 20 features for reduction. The performance of XGBoost exceeded all other techniques based on the combined evaluation of accuracy, recall, precision, and F1-scores, which attained rates of 0.983, 0.984, 0.967, and 0.970 respectively.

TABLE 5.1
Comparison of Various Related Work

Author(s)	Contribution	Model(s)	Feature Selection	Dataset Used	Performance
Gad et al. [16]	Proposed an online ML-based IoT threat detection system	DT, RF, XGBoost, KNN	Chi-square statistics (20 features), SMOTE	TON_IoT	Accuracy: 98.3%
Guo et al. [17]	Developed an IDS framework with stacking ensemble models	CatBoost, Extra Trees, AdaBoost, XGBoost, Stacking Ensembles	Not specified	TON_IoT	MCC: 0.9971 (binary), 0.9909 (multi)
Bhandari et al. [18]	Applied multiple ML models on IIoT datasets	DT, RF, NB, SVM, Gradient Boosting, DNN	Undersampling and oversampling	Edge-IIoTset, IoT-23	Accuracy: 96%
Li et al. [19]	Proposed an anomaly-based IDS with SVM + CNN + RNN layers	SVM with CNN and RNN	NA	TON_IoT	Accuracy: 99%
Ramaiah et al. [20]	Evaluated multiple models for Network IDS	ERT, XGBoost, LSTM, AdaBoost	Z-score normalization, SMOTE	Edge-IIoTset	Accuracy: 99.93%
Hamza et al. [21]	Investigated malware detection with supervised ML techniques	KNN, DT, LR, SVM, RF	NA	Edge-IIoTset	Accuracy: 94%

Guo et al. [17] established an IoT IDS framework that applied TON_IoT data to test ten ML approaches including CatBoost and Extra Trees together with AdaBoost and XGBoost as well as ensemble stacking models. The stacking-ensembled model reached MCC values of 0.9971 and 0.9909 as the best performance in both binary and multiclass classification.

The researchers applied DT, RF, NB, SVM and Gradient Boosting, and DNN to test both Edge-IIoTset and IoT-23 datasets according to Bhandari et al. [18]. They achieved the best outcomes using SVM and DT via accuracy, precision, recall, and F1-score by balancing classes by undersampling and oversampling. For accurate classification, Li et al. [19] suggested an anomaly-based IDS that combines SVM with convolutional and recurrent layers. Their method achieved high detection accuracy for a variety of attack types and showed notable gains on TON_IoT.

Models such as Extremely Randomized Trees (ERT), XGBoost, LSTM, and AdaBoost were evaluated on the EdgeIIoTset by Ramaiah et al. [20]. ERT outperformed other network IDS with an accuracy of 99.93% using Z-score normalization and SMOTE for balancing.

For malware detection on the EdgeIIoTset, Hamza et al. [21] employed supervised ML techniques such as KNN, DT, LR, SVM, and RF. Out of all the evaluated methods, RF had the best accuracy (94%).

5.3 PROPOSED FRAMEWORK

The illustration in Figure 5.1, presents an Edge-IIoT intrusion-detection design that makes use of mobile-edge networking systems to create stronger industrial security measures for IIoT infrastructure. The system initiates from user-end IoT devices such as smartphones and wearable sensors which produce application traffic. The edge layer serves as the main point where real-time data processing and threat mitigation procedures take place after traffic data is received.

The system at the edge implements multiple integrated functionalities that allow it to function.

1. The system monitors network behavior through Distributed Traffic Monitoring while it searches for unusual behavior patterns.
2. The Data Traffic Controller operates as a framework for handling data movement to avoid system bottlenecks when guaranteeing steady communication streams.
3. An Intrusion Detection System utilizes ML models to recognize dangerous system behaviors.

Comprehensive Intrusion Detection through Hybrid Intrusion Detection optimizes both ML-based approach performance and eliminates false-alarm rates through feature-selection techniques.

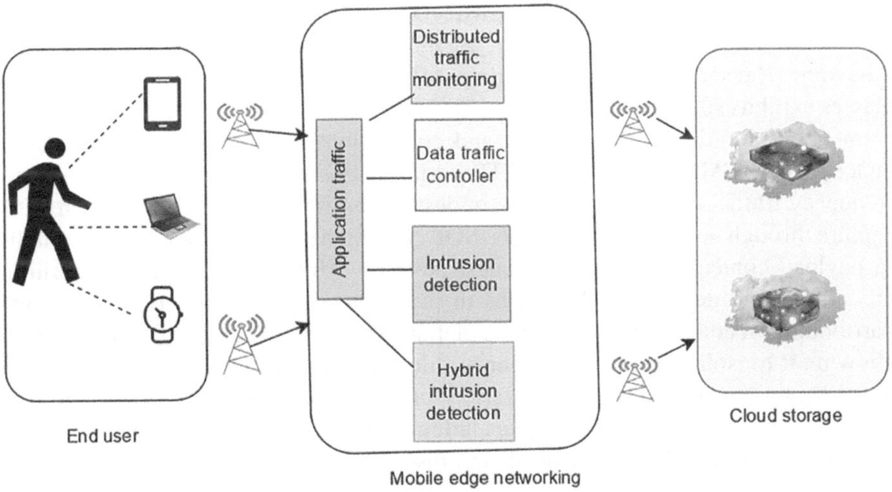

FIGURE 5.1 Illustration of attack detection mechanism in the IoT-Edge-Cloud scenario.

Using the Optuna-feature selection method selects network traffic attributes for the ML model thus improving its performance and operating efficiency. The secured ground data transfers to cloud storage where further analysis can be conducted as well as long-term storage and higher-level decision-making operations take place.

The designed architecture enables immediate edge-based smart-threat identification which results in both reduced network latency and bandwidth consumption as well as powerful protection against new cybersecurity dangers existing in IIoT systems.

5.3.1 Random Forest Classifier

The Random Forest Classifier is an ensemble learning algorithm that builds a collection of decision trees and aggregates their outputs to perform classification. Mathematically, for a classification task with 15 classes $y \in \{1,2,3,\ldots,15\}$, the model constructs T decision trees $(h_t(x))_{t=1}^{T}$, where each tree is trained on a random bootstrap sample of the original training data. At each node in a tree, a random subset of features is selected, and the best split is chosen based on a purity criterion such as Gini impurity, defined as $G = 1 - \sum_{i=1}^{15} p_i^2$, where p_i is the proportion of class i instances at the node. Each decision tree outputs a class prediction $h_t(x) \in \{1,2,3,\ldots,15\}$. The final prediction of the random forest is obtained by majority voting:

$$f(X) = arg \max_{k \in \{1,2,\ldots,15\}} \sum_{t=1}^{T} \delta\left(h_t(x) = k\right)$$

where δ is the indicator function. This approach reduces overfitting by averaging multiple models and increases accuracy by leveraging the diversity among trees due to random sampling and feature selection.

In multi-class IDS, RFC proves effective in classifying complex cyberattack types like the 15 classes: DDoS_ICMP, DDoS_UDP, DDoS_TCP, DDoS_HTTP, XSS, SQL_injection, Backdoor, Port_Scanning, Uploading, Vulnerability_scanner, Password, Ransomware, Fingerprinting, MITM, and normal traffic. Each of these classes exhibits specific statistical patterns in network traffic features—such as packet count, flow duration, protocol type, and connection behavior. For example, DDoS attacks (like ICMP, UDP, TCP, HTTP variants) generally produce bursts of high-frequency traffic with low variability in destination IPs, which RFC trees can easily capture through multiple feature splits. SQL_injection and XSS often show anomalies in payload content and unusual request patterns, while MITM and Fingerprinting are subtler and depend on deviations in protocol interaction patterns. Recursively partitioning feature space at various depths RFC captures these characteristics, allowing it to isolate distinct behavior profiles per class. RFC is a reliable technique for recognizing and categorizing a variety of cyberthreats in IIoT and edge network contexts since it repeatedly aggregates judgments across trees, improving both precision and recall in highly unbalanced circumstances.

5.3.2 Optuna-based Feature Selection

Optuna is a sophisticated and adaptable hyperparameter optimization framework created especially to automate the process of finding the best settings for ML models. The sampling approach of Tree-structured Parzen Estimator (TPE) serves SMBO as an effective tool to reach its goal. The feature enables Optuna to learn from past experiments for better guidance in subsequent searches of the hyperparameter space. Optuna's dynamic search engine constructs spaces on the fly to eliminate unwanted trials before time runs out, thus boosting time efficiency and accuracy of high-performing configurations discovery. Optuna works effectively with diverse optimization tasks because its lightweight structure, its ability to interact with essential ML libraries together with parallel support makes it usable for small and large-scale projects.

The performance of ML models benefits from the considered Optuna technique, which optimizes key hyperparameters to regulate algorithm behavior within Random Forest and Support Vector Machines and Neural Networks. The identification of the most effective hyperparameter combinations for models occurs when Optuna conducts evaluation metrics (accuracy, F1-score) throughout various trials. Optuna implements its trial-based paradigm, which enables researchers to specify objective functions that evaluate model performance through defined evaluation logic and additionally provides automatic experimental management of logging and early-termination features. This results in more robust and generalizable models, enabling researchers and practitioners to deploy ML solutions with improved predictive accuracy and efficiency—especially critical in real time and high-stakes domains such as cybersecurity, healthcare, and industrial automation.

5.4 RESULTS

Experimental Setup

For the experiments, we used a system configured with 16 GB of RAM and a 12th-generation Intel Core i7 processor running at a clock speed of 2.10 GHz. The experimental procedures were implemented using the latest version of Python. Commonly used libraries such as pandas, numpy, scikit-learn, Optuna, and Matplotlib were utilized for data handling, model building, hyperparameter tuning, and result visualization.

Dataset Description

A new comprehensive cybersecurity dataset named Edge-IIoTset [22], from which the ML-EdgeIIoT-dataset is used for our experiment, designed specifically for IoT and Industrial IoT (IIoT) applications. The dataset supports both centralized and federated learning models for ML-based IDS. The testbed used to generate the dataset is structured into seven layers, incorporating modern technologies such as blockchain, cloud computing, fog computing, and edge computing. IoT data were collected from more than ten different sensor types, including temperature, humidity, water level, pH, and heart rate sensors. The dataset includes 14 different cyberattacks, grouped into five main threat categories: DoS/DDoS, information gathering, man-in-the-middle,

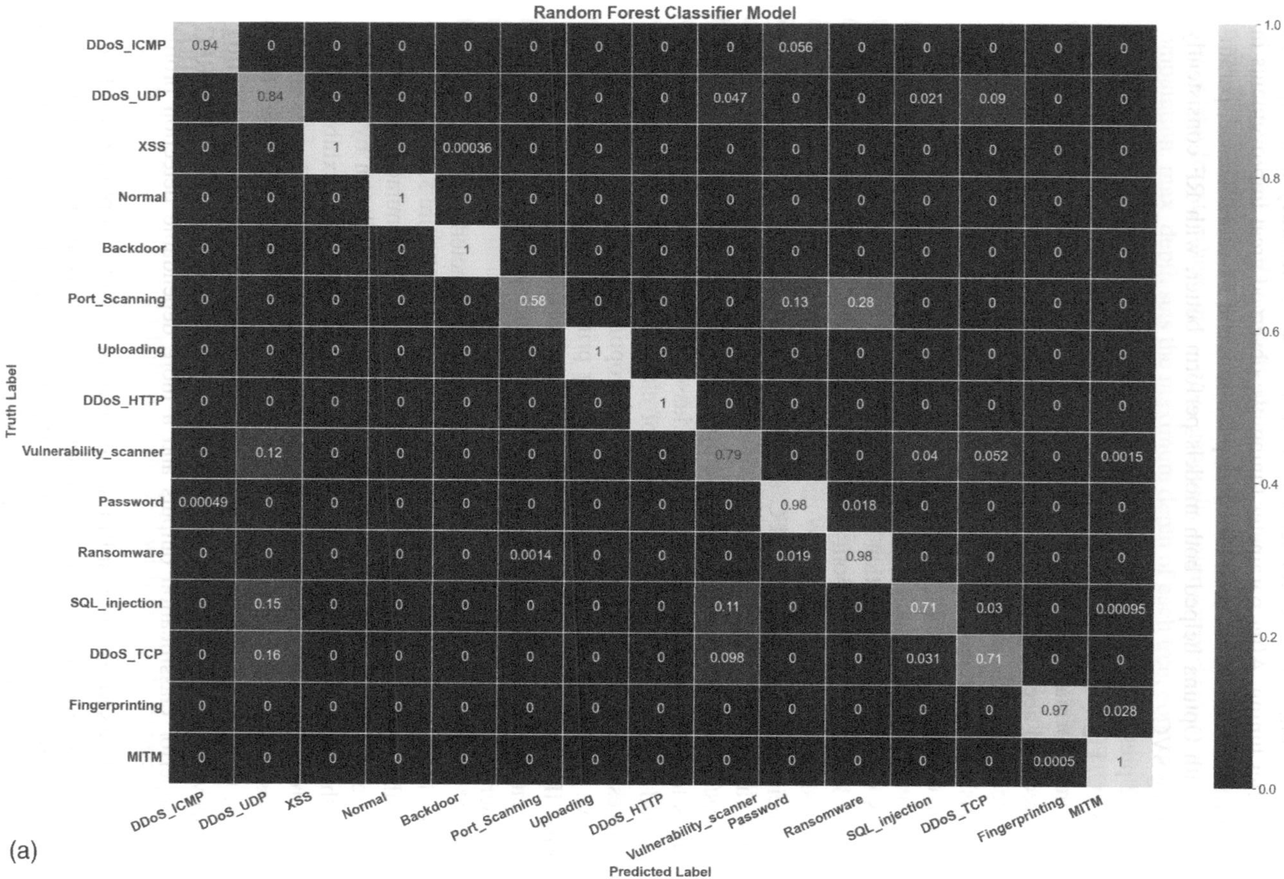

(a)

FIGURE 5.2 Shows the comparison of CM for different classifier. (a) CM for Random Forest Classifer. (b) CM for Support Vector Classifier.

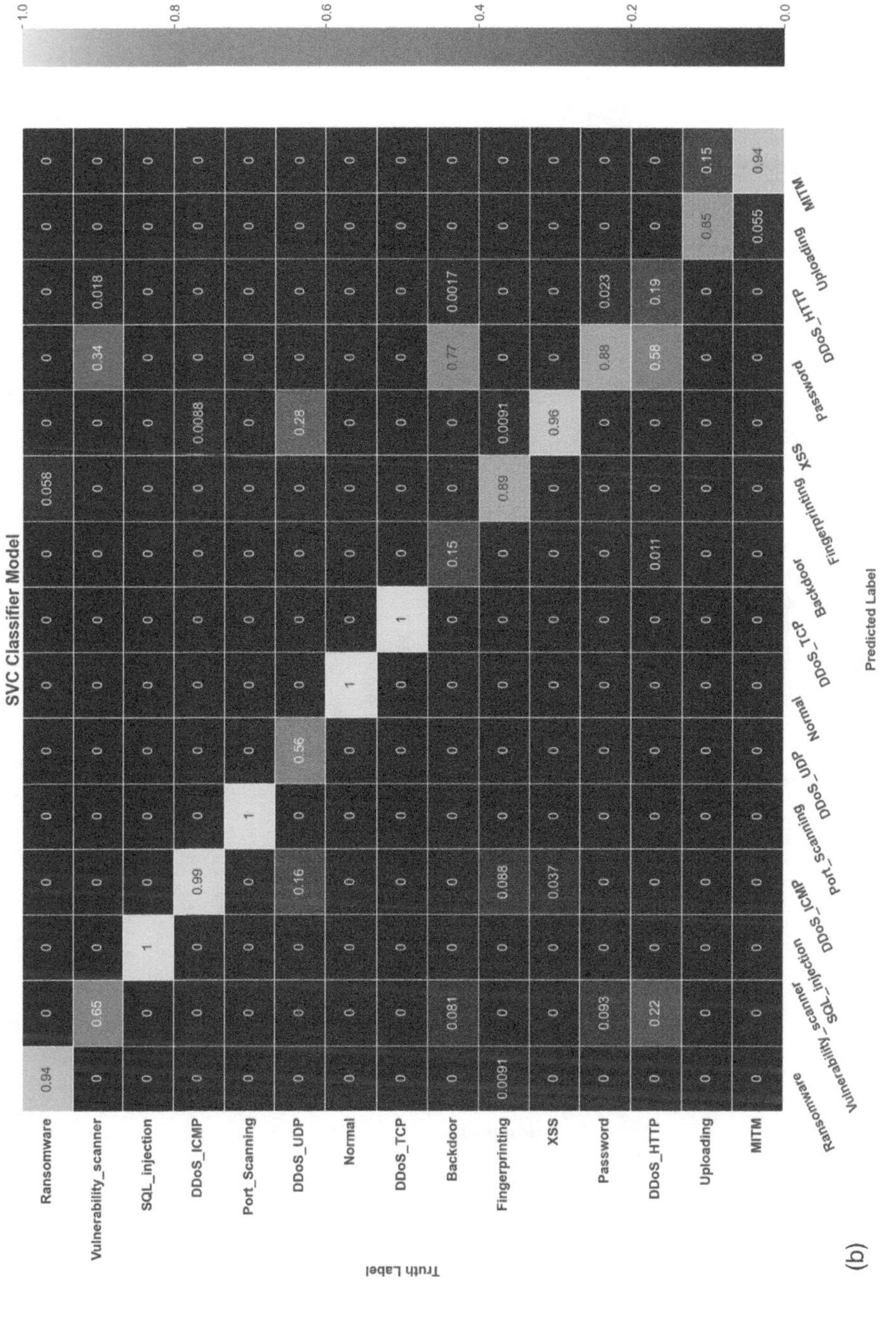

(b)

FIGURE 5.2 (Continued)

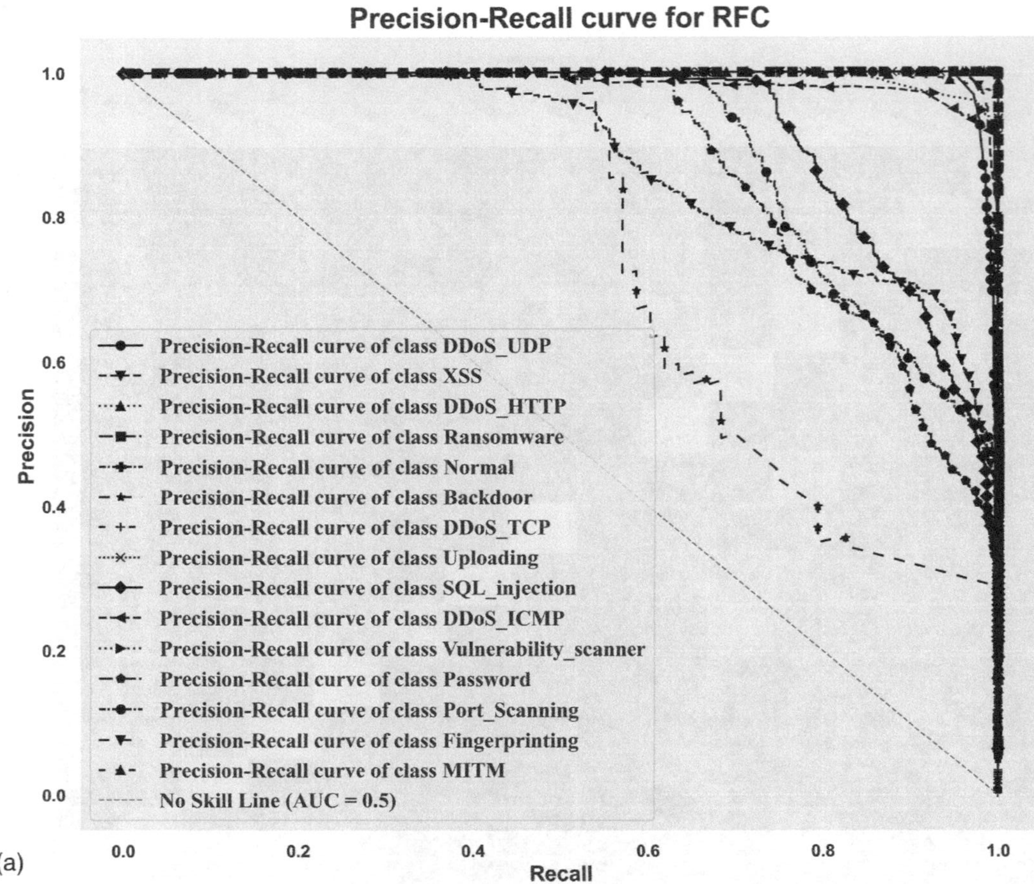

(a)

FIGURE 5.3 Shows the comparison of Precision-Recall Curve for different classifier. (a) Precision-Recall Curve for Random Forest Classifer. (b) Precision-Recall Curve for Support Vector Classifier.

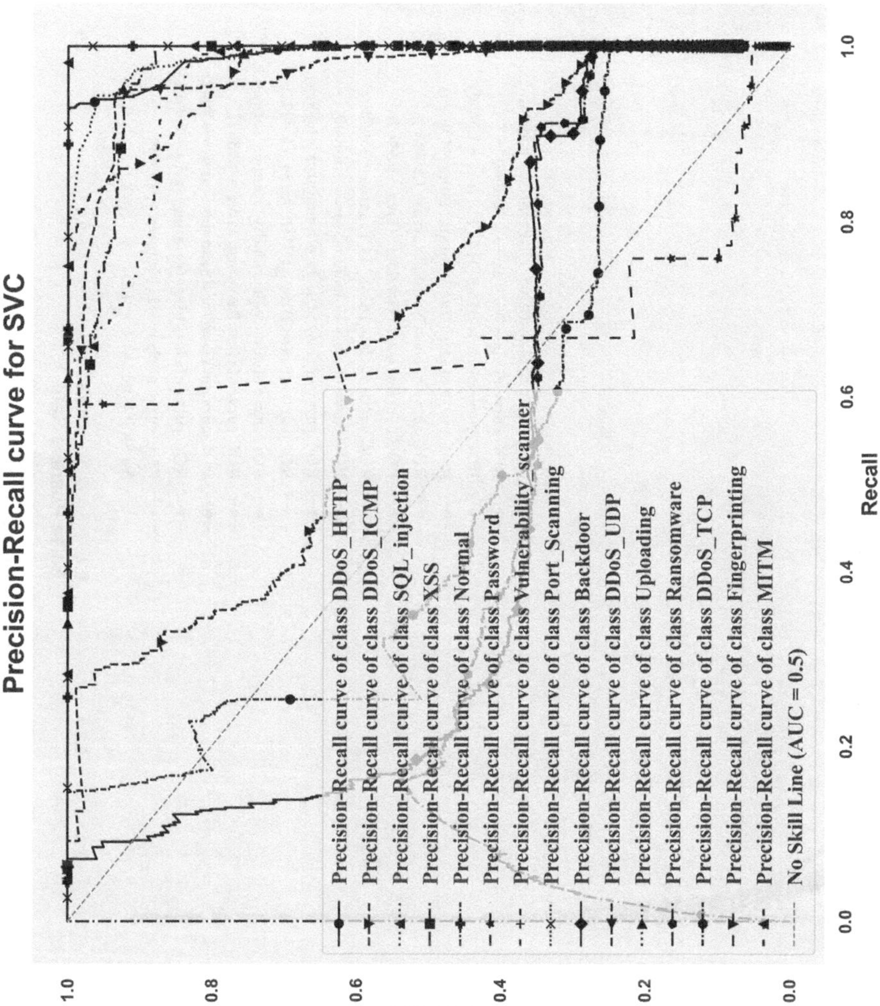

(b)

FIGURE 5.3 (Continued)

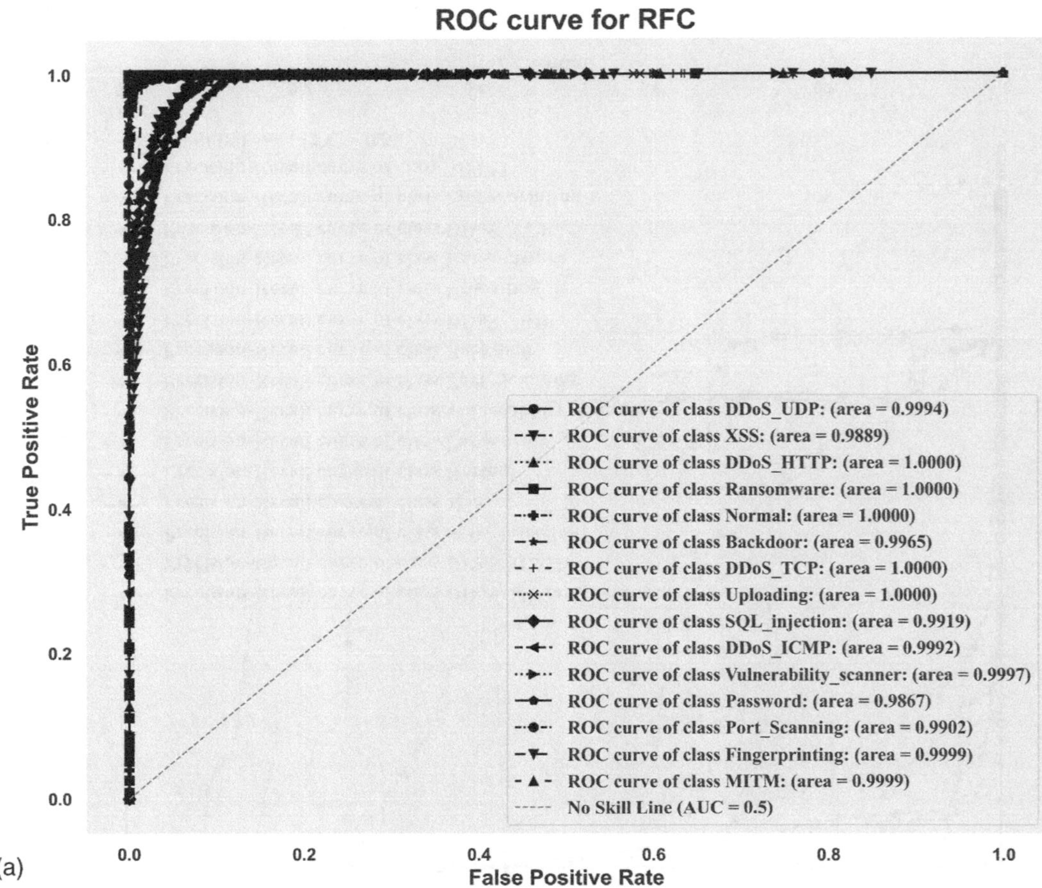

(a)

FIGURE 5.4 Shows the comparison of ROC Curve for different classifier. (a) ROC Curve for Random Forest Classifer. (b): ROC Curve for Support Vector Classifier.

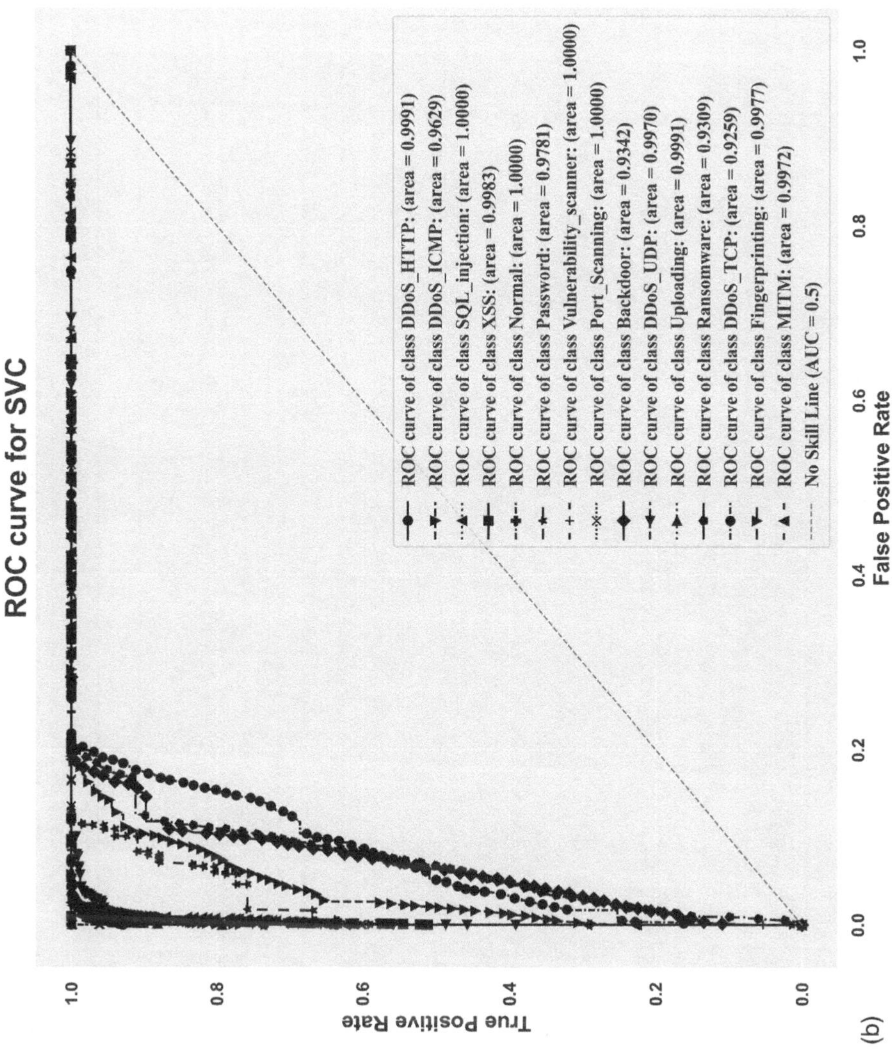

(b)

FIGURE 5.4 (Continued)

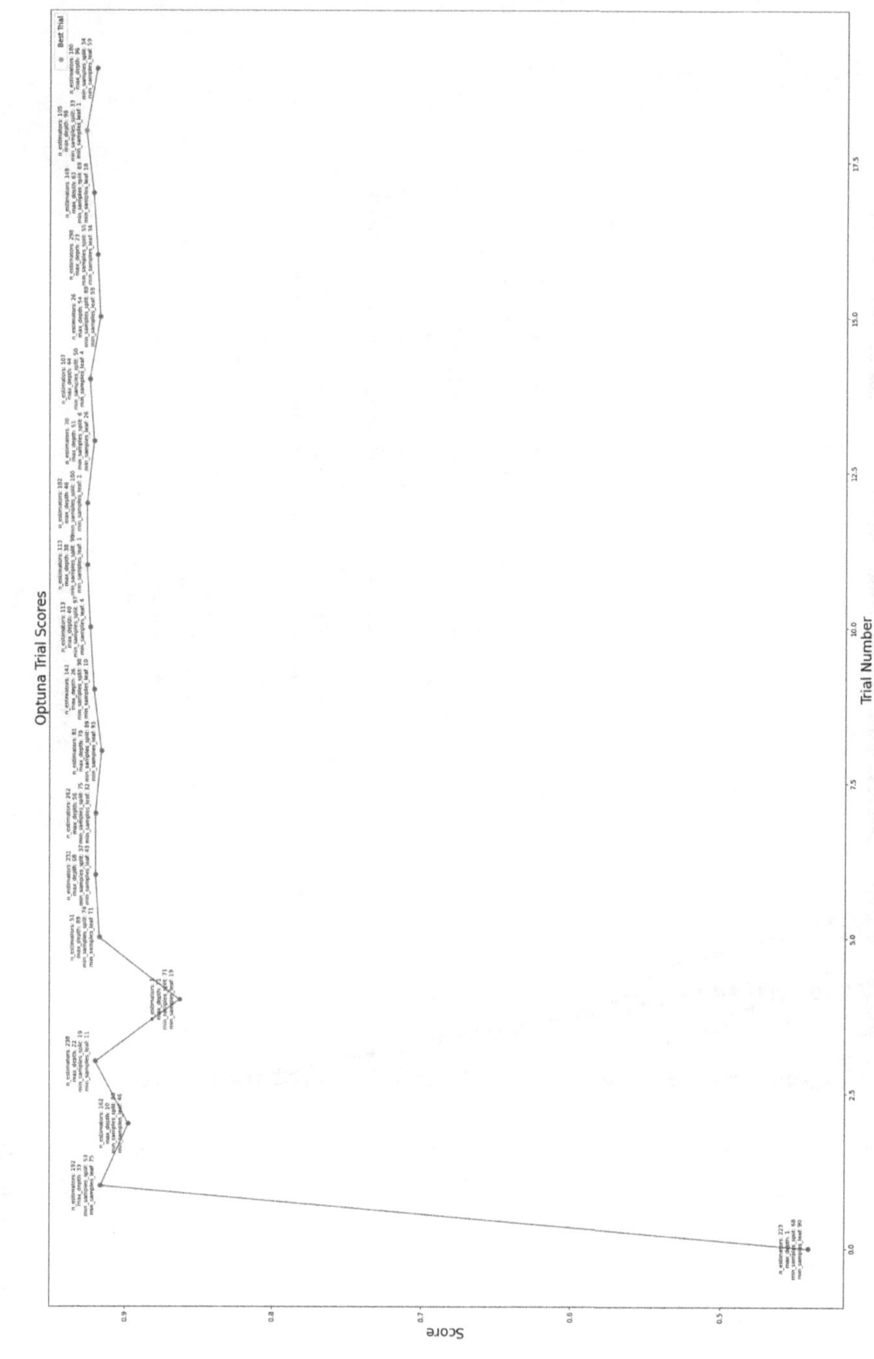

FIGURE 5.5 Shows the scores achieved by the Random Forest Classifier using different parameters across multiple trials using Optuna.

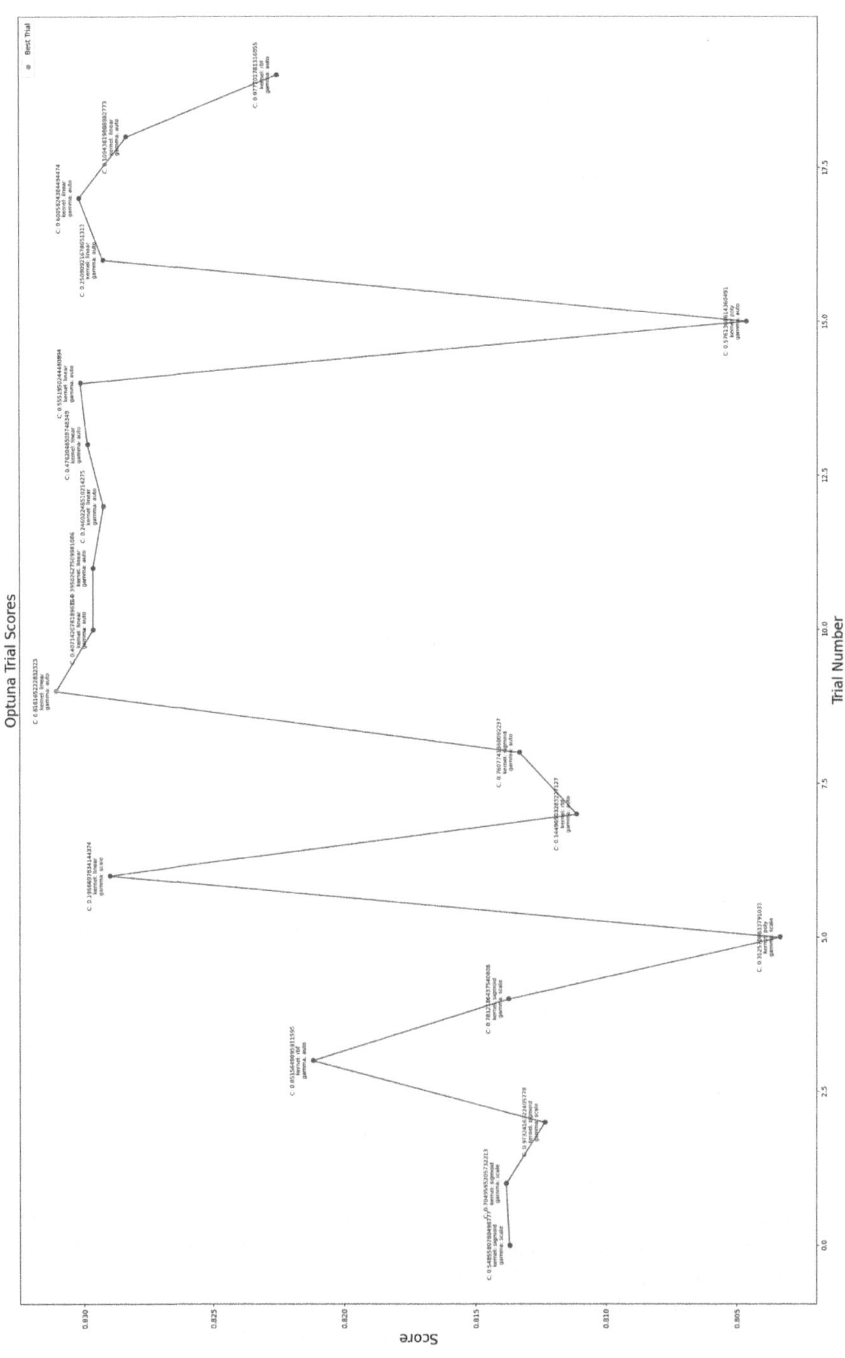

FIGURE 5.6 Shows the scores achieved by the Support Vector Classifier using different parameters across multiple trials using Optuna.

performance ratings for each of the 20 Optuna trials. Different combinations of C, kernel, and gamma values were evaluated to optimize the classifier's accuracy. As the trials progressed, the model demonstrated gradual performance improvements. The best accuracy score achieved was 0.8310, which was obtained using the following parameters: C = 0.8161, kernel = 'linear', and gamma = 'auto'.

5.5 CONCLUSIONS AND FUTURE SCOPE

This chapter establishes an effective intrusion detection solution for IIoT systems through the combination of ML models, which optimizes their hyperparameters using Optuna-based techniques. Real-time threat detection becomes achievable through model deployment at network edge positions utilized as part of this method for time-sensitive industrial applications, which also enables latency reduction. RFC proved to be the best-performing model since it surpassed all its competitors by reaching 92.67% accuracy in only 20 trials. The automated approach to optimization stands crucial in both accuracy development and increased system robustness for IDS. Through the provided framework practitioners acquire a viable method to detect various IIoT network threats while benefiting from scalability and maintaining reliable performance levels.

The research on federated learning techniques for this application field should advance to enable multiple edge nodes to share intrusion detection data while upholding privacy integrity. The implementation of LSTM or CNN DL models provides systems with the ability to detect complex patterns together with time-based patterns in IIoT traffic. The system will become better at detecting emerging threats when it implements real-time threat intelligence feeds for activating automatic model updates. The development of compact models which can run effectively on minimally powered devices stands as a main focus.

REFERENCES

[1] Jha, S., Nkenyereye, L., Joshi, G.P., & Yang, E. (2020). "Mitigating and Monitoring Smart City Using Internet of Things," *CMC-Computers, Materials & Continua*, vol. 65, no. 2, pp. 1059–1079. https://doi.org/10.32604/cmc.2020.011754.

[2] Oks, S. J., Jalowski, M., Lechner, M., Mirschberger, S., Merklein, M., Vogel-Heuser, B., & Möslein, K. M. (2024). Cyber-physical systems in the context of industry 4.0: A review, categorization and outlook. *Information Systems Frontiers*, *26*(5), 1731–1772.

[3] Saranya, T., Sridevi, S., Deisy, C., Chung, T. D., & Khan, M. A. (2020). Performance analysis of machine learning algorithms in intrusion detection system: A review. *Procedia Computer Science*, *171*, 1251–1260.

[4] Ahmad, R., Alsmadi, I., Alhamdani, W., & Tawalbeh, L. A. (2022). A deep learning ensemble approach to detecting unknown network attacks. *Journal of Information Security and Applications*, *67*, 103196.

[5] Jha, S., Tariq, U., Joshi, G. P., & Solanki, V. K. (Eds.). (2022). *Industrial Internet of Things: Technologies, Design, and Applications*. CRC Press.

[6] Shakya, S. R., & Jha, S. (2022). Challenges in Industrial Internet of Things (IIoT). In S. Jha, U. Tariq, G. P. Joshi, & V. K. Solanki (Eds.), *Industrial Internet of*

Things: Technologies, Design, and Applications (pp. 21–40). CRC Press. https://doi.org/10.1201/9781003102267-2

[7] Bebortta, S., & Singh, S. K. (2021, June). An adaptive machine learning-based threat detection framework for industrial communication networks. In 2021 10th IEEE International *Conference* on *Communication Systems* and *Network Technologies* (CSNT) (pp. 527–532). IEEE.

[8] Shakya, S. R., & Jha, S. (2021). Internet of Things application in life sciences. In R. K. Tyagi, M. B. Patil, & T. K. Shah (Eds.), *The Role of Internet of Things (IoT) in Biomedical Engineering: Present Scenario and Future Scope* (Chapter 9). Academic Press. https://doi.org/10.1016/B978-0-12-817356-5.00012-7

[9] Sarhan, M., Layeghy, S., Moustafa, N., Gallagher, M., & Portmann, M. (2024). Feature extraction for machine learning-based intrusion detection in IoT networks. *Digital Communications and Networks*, *10*(1), 205–216.

[10] Konecny, J., Choutka, J., Hercik, R., Koziorek, J., Navikas, D., Andriukaitis, D., & Prauzek, M. (2024). Computational cost and implementation analysis of a wavelet-based edge computing method for energy-harvesting industrial IoT sensors. *IEEE Access.*

[11] Abdullahi, S. M., & Lazarova-Molnar, S. (2025). On the adoption and deployment of secure and privacy-preserving IIoT in smart manufacturing: A comprehensive guide with recent advances. *International Journal of Information Security*, *24*(1), 53.

[12] Tiwari, R. S., Lakshmi, D., Das, T. K., Tripathy, A. K., & Li, K. C. (2024). A lightweight optimized intrusion detection system using machine learning for edge-based IIoT security. *Telecommunication Systems*, 1–20.

[13] Ullah, F., Ullah, S., Srivastava, G., & Lin, J. C. W. (2024). IDS-INT: Intrusion detection system using transformer-based transfer learning for imbalanced network traffic. *Digital Communications and Networks*, *10*(1), 190–204.

[14] Arya, A., & Jha, S. (2022). An analytical review on data privacy and anonymity in "Internet of Things (IoT) enabled services." In 2022 6th International Conference on Computing, Communications and Security (ICCCS*)* (pp. 1–6). IEEE. https://ieeexplore.ieee.org/document/9725770

[15] Shakya, S. R., & Jha, S. (2024). Introduction to Industrial Internet of Things (IIoT) – toward the future internet. In S. R. Shakya & S. Jha (Eds.), *Quality Assessment and Security in Industrial Internet of Things* (pp. 1–19). CRC Press. https://doi.org/10.1201/9781003530572

[16] Gad, A. R., Haggag, M., Nashat, A. A., & Barakat, T. M. (2022). A distributed intrusion detection system using machine learning for IoT based on ToN-IoT dataset. *International Journal of Advanced Computer Science and Applications*, *13*(6).

[17] Poudyal, Y. R., Sharma, G., Adhikari, L. B., Chetry, A. B., & Jha, S. (2025). Developing Internet of Things (IoT) based smart government model for Nepal. *AIP Conference Proceedings*, *3224*(1), 020007. https://doi.org/10.1063/5.0245896

[18] Jha, S., Nkenyereye, L., Joshi, G. P., & Yang, E. (2020). Mitigating and monitoring smart city using Internet of Things. *Computers, Materials & Continua*, *65*(2), 1059–1079. https://doi.org/10.32604/cmc.2020.011754

[19] Li, S., Chai, G., Wang, Y., Zhou, G., Li, Z., Yu, D., & Gao, R. (2023). Crsf: An intrusion detection framework for industrial internet of things based on pretrained cnn2d-rnn and svm. *IEEE Access*, *11*, 92041–92054.

[20] Jha, S., Ahmad, S., Abdeljaber, H. A. M., Othman, N. A., Nazeer, J., & Yadav, K. (2025). Enabling resilient wireless networks: OFDMA-based algorithm for enhanced survivability and privacy in 6G IoT environments. *IEEE.* https://ieeexplore.ieee.org/abstract/document/10445489

[21] Hamza, N., Lakmal, H. K. I. S., Maduranga, M. W. P., & Kathriarachchi, R. P. S. (2023, August). Malware detection of IoT networks using machine learning: An experimental study with Edge IIoT dataset. In 30th Annual Technical Conference-IET Sri Lanka Network, Colombo, Sri Lanka.

[22] Son, L. H., Jha, S., Kumar, R., Chatterjee, J. R., & Khari, M. (2019). Collaborative handshaking approaches between the internet of computing and the internet of things towards a smart world: A review from 2009–2017. *Telecommunication Systems, 70*(4), 617–634. https://doi.org/10.1007/s11235-018-0481-x

6 Machine Learning-Powered Intrusion Detection for Securing IoT and IIoT Networks

Subhranshu Sekhar Tripathy, Sujit Bebortta, Niva Tripathy, Satyananda Champati Rai, and Subhendu Kumar Pani

6.1 INTRODUCTION

The IoT has become more and more widespread in a number of industries, including home appliances, healthcare, and automobiles. IoT technology has transformed our daily lives [1]. The particular application of IoT in the present industry led to the development of the Industrial Internet of Things (IIoT) concept. Today, IIoT refers to the application of traditional IoT across a range of industrial programs and businesses. The IIoT is an extension of the IoT that concentrates on industrial use cases [2]. Vehicle networks, home appliance networks, enhanced security systems, actuators, sensors, control systems, communication and integration interfaces, and more are some of its many constituents [3]. IIoT devices, like sensors on manufacturing equipment or embedded chips in power networks, provide organizations with real-time information that aids in enhancing operational efficiency, optimizing performance, and carrying out predictive maintenance [4]. IIoT nodes can connect to the Internet using a variety of communication protocols, such as cellular, LoRaWAN, TCP/IP, MQTT, and Modbus TCP.

One of the IIoT architectural levels is the perception layer, which deals with physical components. The application layer receives data from the network layer, processes it, and shows it to the user. More storage, scalability, and sophisticated analytics are provided by the cloud layer [5, 6]. In addition, the majority of IIoT nodes are capable of collecting, analyzing, and transmitting data. The unique features of IIoT systems and applications may expose them to privacy and security concerns [7]. IIoT systems experience various possible attacks along with intrusive attempts throughout their entire operational structure. Different types of intrusions combined with assorted attacks endanger IoT security through various methods, such as jamming and Denial of Service (DoS), together with spoofing and data manipulation and access control and operating system (OS) attacks. Businesses deploy Intrusion Detection Systems (IDSs) for protecting their IIoT nodes from hostile attacks while maintaining network security [3]. The IDS safeguards IIoT networking by ensuring integrity protection

DOI: 10.1201/9781003609773-6

along with confidentiality and network security. IDS implementation effectively identifies threats that target IIoT networks as it detects threats and takes necessary countermeasures before these harmful activities occur [8]. Anomaly-based and hybrid systems along with signature-based systems constitute the three main IDS system categories. Anomaly-based IDS detects system irregularities differently from signature-based IDS which depends on attack signatures acquired from their databases. The three traditional IDS categories include signature-based systems and hybrid systems and anomaly-based systems. The traditional IDS experiences numerous difficulties that lead to excessive erroneous alerts and restricted detection ability. Additionally, they are unable to identify new types of intrusions or defend against zero-day attacks. Studies have been conducted to determine if artificial intelligence (AI) may enhance the performance of conventional IDS [9, 10].

Machine learning (ML), a branch of AI, has lessened the demand for human involvement in several fields. The methodology has two stages: training, which entails creating a trained model by identifying patterns in data. The testing phase is determined by the model's ability to recognize new cases. The two types of ML are supervised and unsupervised. ML can identify abnormal and dangerous information by training detection models on both benign and malicious IoT data [11]. Several ML techniques are used to detect dangerous data. These methods assume that training data comes from similar sources and kinds (such as pixels) and is homogeneous. Most IoT systems today use diverse data formats such as images, text, graphs, streams, and time series, which are not designed to function with their inherent properties [12–14]. Supervised ML uses labeled datasets to improve decision-making and predict future outcomes. Unsupervised ML is used for learning tasks with unlabelled data. Semi-supervised ML systems learn from both labeled and unlabelled data. Depending on interactions in a particular environment, reinforcement ML determines rewards or errors [15–17]. Decision trees (DT), K-Nearest Neighbors (KNN), Naïve Bayes (NB), support vector machines (SVM), random forests (RF), and artificial neural networks (ANN) are among the supervised classifiers used in IDS design. Unsupervised learning trains models using unlabeled data. IDS employs a number of unsupervised techniques, such as IF, DBSCAN, and k-means. A popular tool for supervised learning and prediction is the RF classifier.

Our research proposes an IIoT IDS based on tree-based supervised ML methods. ML-based IDSs often train on the most recent datasets. Most current datasets are big, both in terms of feature space and network traces. Having too many features in a dataset might hinder training of ML algorithms. ML approaches typically perform worse with more features. Increasing the amount of attributes in a dataset makes learning more difficult [18]. One fundamental advance is the transformation of data from IoT and IIoT devices into images, which makes it possible to employ reliable convolution neural networks (CNNs) for anomaly identification. Pattern recognition and feature extraction are much improved by this new format. To improve accuracy and robustness, we employ an ensemble approach that combines multiple pre-trained CNN architectures that have been adjusted using optimal hyperparameters. Furthermore, we address the issue of resource limitations in IoT systems by minimizing computational load by model quantization while maintaining system performance. Through extensive testing, we show that our suggested method outperforms the current methods in detecting different types of cyberattacks.

Our contributions to this chapter are summarized below:

- Developing an IIoT anomaly detection framework to better detect malicious data from compromised devices.
- Proposing an effective approach for extracting robust IIoT-based features while removing superfluous ones. This leads to improved anomaly detection performance in the IIoT.
- Using actual IoT datasets to investigate and evaluate the proposed method's efficacy. For the purpose of identifying IIoT abnormalities, we contrast our method with current deep learning techniques.

The rest sections of this chapter are organized as follows. Section 6.2 covers the relevant work. Section 6.3 outlines the proposed framework. In Section 6.4, we present our empirical findings. Section 6.5 finishes the chapter.

6.2 RELATED WORK

This section covers recent ML and deep learning algorithms for detecting anomalies in IoT and IIoT network. Many studies have explored using ML to detect anomalies. This section summarizes related studies on IDS utilizing ML approaches. It provides a survey of existing IDS frameworks and methods for detecting intrusions in IoT systems.

The pretraining Wasserstein generative adversarial network IDS (PWG-IDS) was created by Zhang et al. [19]. The researchers double-trained their model using Wasserstein's generative adversarial network with gradient penalty (WGAN-GP) after using LightGBM to detect intrusions in IIoT networks. IDS achieved a 99% F1 score and 100% accuracy on the NSL-KDD dataset, while it achieved a 90% F1 score and 96% accuracy on the CIC-IDS2018 dataset. Using Particle Swarm Optimization (PSO) for feature selection and the Support Vector Machine (SVM) ML technique for classification, Liu et al. [5] created an IoT IDS system. LightGBM functions as the PSO method for this research study. Performance indicators accuracy along with False Alarm Rate (FAR) were used by the writers when they tested their model against the UNSW-NB15 dataset. According to Saxe and Berlin [20] the PSO-LightGBM achieved 86.68% overall accuracy, but with a high 10.62% FAR for intrusion detection. These rectified linear units (ReLU) activation functions provide DNN models with substantial improvements in their ability to detect anomalies among other activation function approaches. A Bayesian model served the framework through its capability to verify whether reported activities should carry the status of real intrusions. The deployment of IDS for IoT networks through ANN was presented in research by Hanif et al. [21]. The system functioned to solve security problems that existed in Internet of Things (IoT) networks. An ML-based IDS system should act as the primary security mechanism due to its ability to deal with the inadequate processing abilities of IoT devices according to the recommendation from the author. The UNSW-NB15 was used to test the proposed method.

In [22] presented a two-stage (TS) paradigm for IDS. The methodology involved detecting minority incursions first, followed by majority attacks in the second step. This paper employs the RF approach for ML classification. Information Gain (IG) was used by the authors to extract features. The UNSW-NB15 dataset was used to assess the IG-TS IDS. FAR and accuracy were the study's parameters. The authors' primary setup in their tests was a binary classification technique. According to the experimental findings, the IG-TS obtained an accuracy score of 85.78% and a FAR of 15.64%. In subsequent research, the writers changing the classifier that was employed in both phases was the aimIn order to improve the security and privacy of IoT networks, Rashid et al. introduced a Federated Learning technique for identifying undesired intrusions while protecting data privacy. Unlike standard centralized ML approaches that require large volumes of centralized data, the proposed FL method trains models locally on IoT devices, sharing just parameter changes with a central server. This server distributes the devices for further training using the enhanced detection method after combining the updates and refining the detection strategy. This distributed method protects privacy while optimizing the general model. The method's effectiveness is demonstrated by extensive experiments on a new dataset, Edge-IIoTset; its accuracy of 92.49% is comparable to 93.92% accuracy of regular centralized ML models [23]. In [24], the authors used the UNSW-NB15 dataset to create an SVM-based NIDS. In order to satisfy the unique needs of IoT networks, this solution was developed. The authors determined that the most important performance metrics were accuracy, detection rate, and false positive rate. Both binary and multi-class classification methods were used in the experiments. The SVM-NIDS completed the binary modeling task with an accuracy of 85.99%. The accuracy of SVM-NIDS was 75.77% in a variety of class scenarios.

An anomaly-based IDS was proposed by Alanazi et al. [20] to lower the risk of cyberattacks on IoT networks. The authors suggested feature selection to find the best features and ensemble learning techniques for detection. 99.984% accuracy, 99.982% precision, 99.984% recall, and 99.983% F1 score were obtained from the experiment on a real-time dataset.

6.3 PROPOSED MODEL

The proposed model introduced in this research is a novel ML-based architecture titled Deep Extreme Learning Machine (Deep-ELM). It is designed with a focus on effective, scalable, and accurate intrusion detection in cloud computing environments—a domain increasingly vulnerable to sophisticated and distributed cyberattacks. Deep-ELM efficiently unites Extreme Learning Machines (ELM) speed and simplicity with deep learning capability to create a performance-efficient computational method.

6.3.1 Overview

As cloud computing spreads among many sectors of economy, guaranteeing network security has become a top priority. Characteristics of cloud environments include

[25] being by nature dynamic, distributed, and scalable—traits that, though good for performance and flexibility, also create great security risks. Sometimes, traditional security methods have difficulty keeping up with the abundance, speed, and variety of cloud-based data. This has called for the creation of astucious, flexible and high-performance IDS viable in demanding surroundings. The Deep-ELM suggested in this research aims to fulfill these demands by integrating the fast training benefits of ELM with the deep feature extraction capabilities of multilayer neural networks. The hierarchical structure enhancement of this threat detection method increases accuracy for known and unknown threats making it move beyond traditional shallow learning methods with Deep-ELM [6]. These algorithms normally need significant time for training together with manual feature selection and automatic parameter adjustments. Cloud-based environments with fast threat vectors and real-time threat detection expose such errors most prominently. The Deep-ELM architecture filters complex patterns across numerous hidden layers through deep learning to solve these problems while ELM speed remains natural thanks to its analytic output calculation using randomly created input weights. Through this advanced design model exhibits stronger data generalization skill while it recognizes fine traffic conduct modifications which results in minimized false positive and negative alarm rates. A distributed deployment plan reflects real cloud computing situations by spreading the data across various spatially distributed nodes to prototype the proposed model. The deployment approach provides both increased scalability along with reduced system latency which normally affects central servers. This autonomous Deep-ELM model system lets its nodes work individually to create a distributed intrusion detection framework that brings decentralized intrusion threat detection through real-time monitoring. Deep-ELM emerges as a small yet adaptable and highly precise IDS developed to fulfill the distinctive needs of cloud infrastructure requirements according to the presented research. The proposed method establishes a new benchmark for IDS performance which integrates classical ELM with deep hierarchical architecture and distributes operations across a cloud framework.

6.3.2 Architectural Design

The Deep-ELM applies a combination of high-speed training from ELM with deep learning architectures for their hierarchical representation abilities (Figure 6.1). The system specifically aims to process high-dimensional nonlinear heterogeneous network traffic data which IDS applications encounter in cloud environments under real-time operational conditions.

6.3.2.1 Core ELM Principles

Deep-ELM is based on the standard ELM framework, a single-hidden layer feed-forward neural network (SLFN). ELM is renowned for its unusual training technique: it assigns input weights and biases randomly and calculates output weights analytically using the Moore-Penrose pseudo-inverse. As it does not rely on iterative adjustment of parameters via backpropagation, which is computationally expensive and sensitive to local minima, this makes ELM really quick to learn. Standard ELM has little feature abstraction because of its shallow architecture notwithstanding these

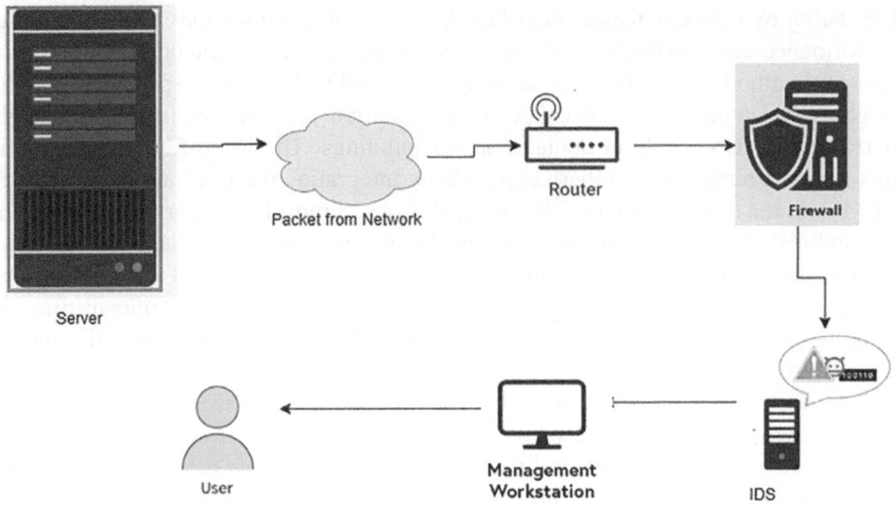

FIGURE 6.1 Architecture of Intrusion detection in cloud computing environments.

benefits. In difficult classification cases where the feature space is very nonlinear—say, separating complex cyberattacks from natural traffic—it could underperform.

6.3.2.2 Deep-ELM Architecture

The suggested Deep-ELM overcomes this shortcoming by organizing several levels of ELMs in a deep and hierarchical fashion. This structure, though meant for ELM's fast learning method, is founded on deep belief networks and stacked auto encoders. The input feature space transforms with every layer operation in Deep-ELM resulting in advanced representations for previous layer outputs. These ELM subnetwork levels obtain training sequentially to produce the ultimate feature abstraction pipeline when considered as independent networks. A last regression or classification component receives the final hidden layer output for supervised training. Deep ELM provides multiple innovative concepts to boost both performance and usability in intrusion detection tasks through its framework. Quick parallelized learning becomes possible through layer-wise training because it functions on every hidden layer separately. The complete backpropagation process becomes unnecessary which leads to simpler training procedures together with faster computations. Deep-ELM utilizes typical ELM noniterative weighting to determine weights between last hidden layer and output layers through Moore-Penrose pseudo-inverse calculations. The computational costs decrease significantly because this method provides fast convergence with no impact on accuracy levels. Deep-ELM features an important strength from its power to build hierarchical feature representations where advanced abstractive capabilities become available one layer at a time from raw input data. Through this learning method the model identifies hidden complex network traffic patterns identifying potential signs of advanced security intrusions. The Deep-ELM approach provides clear paths for network complexity scaling as well as distributed training scalability, which makes it suitable for use in cloud-based IDSs. The flexible structure of Deep-ELM enables its performance in various cyber-physical infrastructures undergoing change.

6.3.2.3 Activation and Regularization

Every ELM layer in the Deep-ELM architecture usually includes nonlinear activation functions like the sigmoid, radial basis function (RBF), or ReLU to introduce nonlinearity into the model. These properties improve the model's ability to capture sophisticated relationships among input characteristics and to learn complex mappings. Analytically computed output weights would cause overfitting, a frequent problem with deep models; therefore regularization methods would be applied. This covers L2 Regularization (Ridge Regression) to penalize great weights.

6.3.2.4 Output Layer

After feature abstraction through multiple ELM layers, the final layer acts as a supervised classifier, trained to distinguish between multiple classes such as different types of attacks and normal traffic. The training target labels are mapped to a one-hot encoded format in the case of multi-class classification tasks. The model can also be adapted for binary classification by using threshold-based decision functions at the output.

6.3.3 Implementation in a Cloud

The deployment strategy for the proposed Deep-ELM model is rooted in the architecture of moderncloud computing environments, which are inherently distributed, scalable, and heterogeneous. Given that cloud infrastructures support a wide array of virtualized services, users, and applications—often across geographically dispersed data centers intrusion detection mechanisms must be equally distributed and adaptable to meet the demands ofreal-time threat monitoring.

To simulate real-world application, the Deep-ELM model was integrated into a distributed intrusiondetection system (D-IDS) framework. In this approach, individual cloud nodes operate as autonomous detection agents, each responsible for monitoring and classifying the network traffic flowing through their respective segments. These nodes are equipped with local instances of the Deep-ELM model, which enables on-site processing of data without the need for centralized coordination. This decentralized strategy offers several advantages, including:

- **Reduced latency** in detection response,
- **Improved fault tolerance** through node-level autonomy,
- **Scalability**, as new nodes can easily join the system and deploy the detection model.

6.3.3.1 Data Flow and Detection Pipeline

The data processing at each node starts with packet capture and preprocessing, where raw traffic data is mapped into NSL-KDD format compatible feature vectors.

The feature vectors then proceed through the Deep-ELM pipeline wherein multiple hidden layers detect intricate patterns corresponding to normal or malicious activities. The final output layer determines each input as a predefined class of normal, DoS, Probe, U2R, or R2L attacks. It enables parallel detection operations, and the system remains very efficient even with high network loads. Additionally, regular synchronization among nodes can be implemented to provide consistency in model parameters and sharing of threat intelligence without sacrificing local autonomy of detection units.

6.3.3.2 Deployment Environment

The simulated cloud environment created using virtual machines (VMs) and containerized instances employed the Deep-ELM model to mimic nodes in a cloud infrastructure. The deployment platform operated under a Windows 10 OS running Python 3.6; every node had 8.00GB of RAM and an Intel Core i7-6500U CPU operating at 2.50GHz. The hardware configuration here supports the model's performance on moderately powered machines and corresponds to typical resources located in the cloud. In reality, Deep-ELM is light enough to be installed on central cloud servers as well as edge computing nodes, thus allowing a hybrid intrusion detection mechanism where data can be analyzed and observed at the network's core as well as its edges. Network security requires this functionality most importantly for situations that involve IoT and mobile edge computing implementations. A local inference process at each node helps keep network sensitive information private due to its dedicated operation without sharing network data between its components. Shephard and Dong (2016) demonstrate that multi-tenant cloud architectures need this design to conform to cloud governance rules by maintaining strict data isolation between different customer groups [17].

6.3.4 Proposed Algorithm

Algorithm 1: Deep-ELM for Intrusion Detection

Require: Training dataset $D=\{(x_i,y_i)\}_{i=1}^{N}$, number of hidden layers L, neurons per layer H, activation function $g(\cdot)$

Ensure: Trained Deep-ELM model, predicted labels for test data **Initialize**

1. input weights $W^{(1)}$ and biases $b^{(1)}$ randomly for the first hidden layer
2. Set L (number of hidden layers), H (neurons per layer), and choose $g(\cdot)$ (e.g., sigmoid, ReLU)

Preprocess data:

3. Normalize input features x_i
4. Split dataset into training and test subsets
5. **For** $l = 1$ *to* L **do**
6. Compute hidden layer output: $H^{(l)} = g(W^{(l)}X^{(l-1)} + b^{(l)})$
7. Set $X^{(l)} = H^{(l)}$
8. Randomly initialize $W^{(l+1)}$, $b^{(l+1)}$ for the next layer
9. **end for**

Compute output weights:
10. Let $H^{(L)}$ be the final hidden output
11. Calculate $\beta = (H^{(L)})^{\dagger} Y$ using Moore–Penrose pseudo-inverse
Prediction phase:
12. For each test sample $\mathbf{x_t}$ do
13. Forward propagate through layers:
$h^{(1)} = g(W^{(1)}x_t + b^{(1)})$
$h^{(2)} = g(W^{(2)}h^{(1)} + b^{(2)})$
... up to $h^{(L)}$
14. Compute output: $\hat{y}_t = h^{(L)} \cdot \beta$
15. Assign label: label = argmax($\hat{y}_t$)
16. **end for**
Deployment in cloud nodes:
17. Deploy trained model on distributed nodes
18. Perform local traffic monitoring and classification
19. Aggregate threat alerts if needed

The suggested Deep-ELM algorithm runs in a systematic way that guarantees good intrusion detection and effective training in a distributed cloud or edge setting. The primary objective is to produce a trained Deep-ELM model that can correctly classify unfamiliar network traffic data starting with a labeled D hidden layer, number of neurons per cycle H, and the activation function g . The first step of the algorithm is the initialization of model parameters; first hidden layer input weights and biases are arbitrarily selected. ELM-learning models depend on this random assignment, which eliminates the need for consistent weight adjustment. The network's architecture is ultimately decided, including the number of layers and the activation function (such as sigmoid or ReLU), thus introducing nonlinearity into the learning process. Normalizing input feature vectors to a constant range in the preprocessing stage guarantees numerical stability and improves the model's capacity to generalize. After that, the record is split into testing and training sets to provide the model's performance independent verification. Training takes place on a layer-wise basis, whereby every hidden layer processes the output of the one before it. The activation function is applied at every layer to change the input and let the model extract increasingly abstract and high-level properties. This method is free of backpropagation and offers quick, parallelizable training.

The algorithm calculates the output weights analytically using the Moore-Penrose pseudo-inverse once the output of the last hidden layer is known. Compared to conventional neural network training techniques, this closed-form answer speeds convergence and dramatically cuts computational overhead. Every test sample is run through the trained model, layer by layer, in the prediction stage until the last output is computed. Choosing the output neuron with the greatest activation value sets the forecasted class label. The trained Deep-ELM model is eventually distributed over edge nodes or distributed cloud. Every node locally monitors and ranks traffic, so offering real-time detection features. Deployment's decentralized structure let for the aggregation of threat warnings if more general network-level analysis or response is required, therefore ensuring flexibility and sensitivity as well as scalability. In scenarios demanding quick, precise, and lightweight intrusion detection, the Deep-ELM approach is particularly well suited for dynamic, large-scale settings.

6.4 RESULTS AND DISCUSSIONS

The testing procedures in this research project operated inside a Python 3.6 platform on Windows 10 with an Intel Core i7-6500U CPU functioning at 2.50 GHz speed alongside 8.00 GB RAM. The NSL-KDD dataset served as the main experimental resource because researchers face significant obstacles when acquiring real-time network attack data from sensing nodes. The Advanced Research and Design Bureau under U.S. Department of Defense originally compiled the NSL-KDD dataset during their intrusion detection project at the MIT Lincoln Laboratory. The dataset includes specific records received from simulated realistic attacks that appear in network connections and platform audit logs. Data scientists have recognized this evaluation tool as their standard benchmark for network intrusion detection studies because it was purpose-built to analyze network data under different circumstances. The main goal of this work involves using IDS data for classifying and predicting network behaviors. The performance assessment of proposed methods makes use of accuracy together with false positive rate (FPR) and false negative rate (FNR) as evaluation metrics.

The research tested three ML techniques including RF, ELM, and the proposed Deep-ELM for intrusion detection in a cloud computing framework. For real-world deployment simulation the Distributed cloud nodes implemented both ELM and Deep-ELM models yet RF acted as the benchmark ensemble model. The NSL-KDD dataset was used for training where 30,000 samples were utilized before dedicating 10,000 samples for testing in the comparative analysis. Multiple experiments were conducted twenty times individually, and the reported data integrated results from all iterations to validate statistics and reduce random variations. Two main performance measurement criteria were employed during evaluation including detection accuracy and training time. Deep-ELM included a hierarchical networking design that used multiple hidden layers to boost feature abstraction capabilities yet ELM required both single hidden layers and analytical weight definitions for output delivery. The RF classifier required an appropriate tree number count to achieve performance and computational efficiency equilibrium. Experimental results demonstrated that Deep-ELM achieved superior accuracy than both ELM and RF yet produced detection accuracy together with robust generalization capability. Deep-ELM demonstrated longer training times than ELM, yet its superior performance made the increased computation acceptable. Deep-ELM demonstrated higher training times and better precision in addition to its reasonable accuracy rate compared to the RF algorithm. The research shows that Deep-ELM proves effective as an intrusion detection tool for cloud systems because it delivers optimal performance and processing speed.

In Figure 6.2, the boxplot reveals that the numerical features for the training set has different distributions based on network traffic class, which shows substantial variation between median points and interquartile ranges across categories. The distribution centers of portsweep and teardrop and nmap classes display larger median metrics and dispersion compared to neptune and normal classes. The boxplot in Figure 6.3, shows how specific numerical data points group across test set network classes relating their central measures and variability levels. The medians and spread of portsweep and teardrop and nmap traffic are higher than normal and neptune and smurf which have narrower spread and clustered behavior.

In Figure 6.4, three algorithms including RF, ELM and Deep-ELM are evaluated toward the accuracy plot for their performance as training data sample volumes increase. The proposed Deep-ELM delivers better accuracy results than RF and ELM because it demonstrates stronger learning abilities and a robust performance especially when working with large data volumes. The accuracy plot in Figure 6.5, shows that Deep-ELM maintains an overall 95% test accuracy throughout all data volume trials which exceeds both ELM and RF results. The accuracy patterns for ELM and RF demonstrate unstable performance and reduced accuracy while Deep-ELM maintains robust behavior with steadily high accuracy rates when dealing with new data entries.

The plot in Figure 6.6 demonstrates how the packet delivery ratio functionality changes when applying RF and ELM models in addition to the proposed Deep ELM technique for different network node numbers. Data transmission robustness remains high for the proposed Deep-ELM model because it produces stable packet delivery ratio values that surpass the results of other models during transmission across various network proportions. Figure 6.7 shows how power consumption changes for RF, ELM and Deep-ELM methods while network node numbers grow from 0 to 20. Through experimentations, it was inferred that Deep-ELM uses the least amount of power at every network node count which establishes its position as an efficient power-saving neural methodology.

6.5 CONCLUSIONS AND FUTURE WORK

The Deep-ELM algorithm proves better than RF and ELM models for IIoT intrusion detection through its evaluation on the NSL-KDD dataset. The solution proves itself as a promising IIoT deployment option due to its stable accuracy range while maintaining high delivery ratio and low energy usage in cloud networks. Deep-ELM demonstrates strong capabilities for real-time security monitoring across industrial networks since it handles both known threats and unknown ones effectively.

Future applications of Deep-ELM in real-world IIoT systems should include studies for adapting it to practical edge devices operating with limited processing capability. Federated learning needs to be integrated and studied because it will preserve accuracy in distributed network detection while ensuring data privacy.

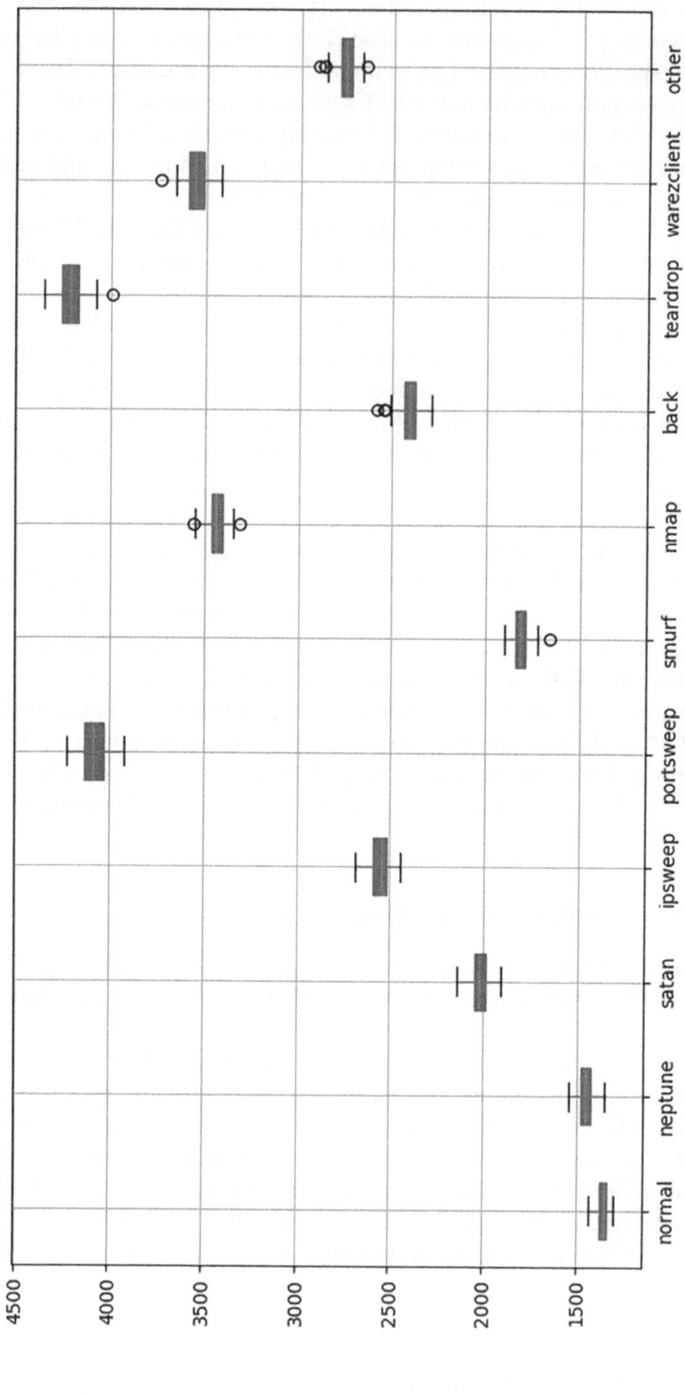

FIGURE 6.2 Visualization of training set's count for attack features.

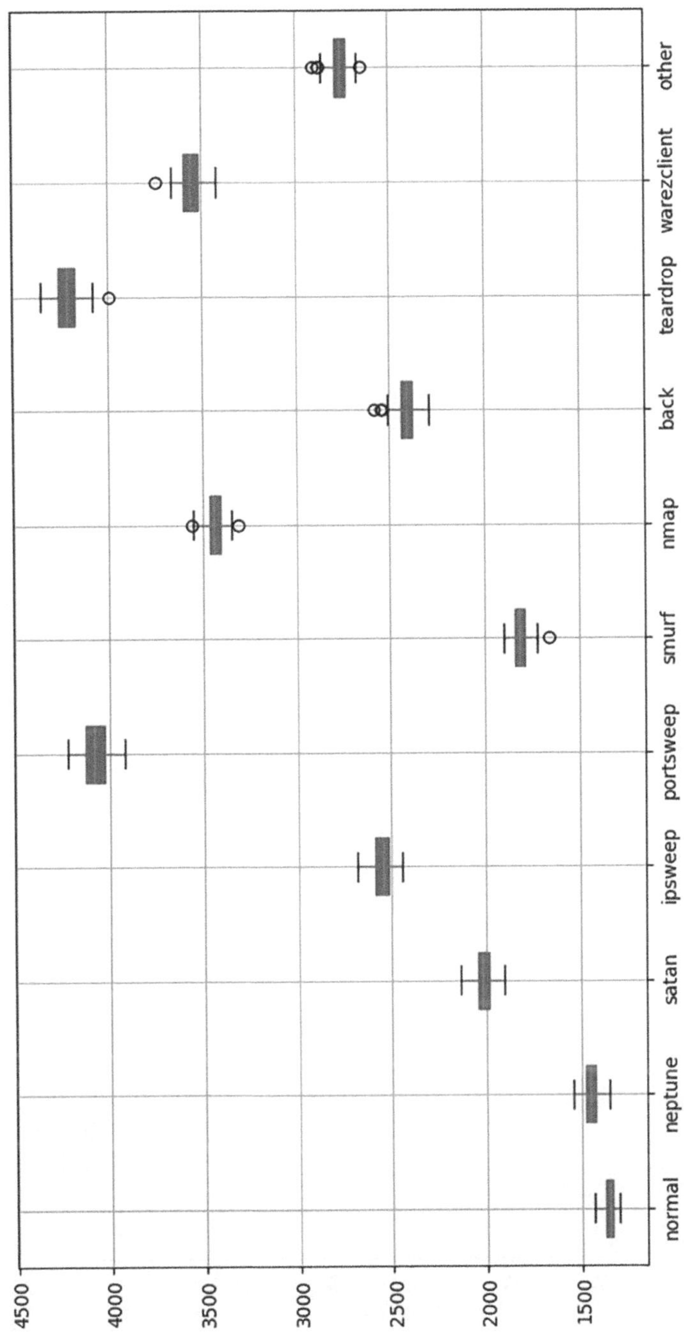

FIGURE 6.3 Visualization of test set's count for attack features.

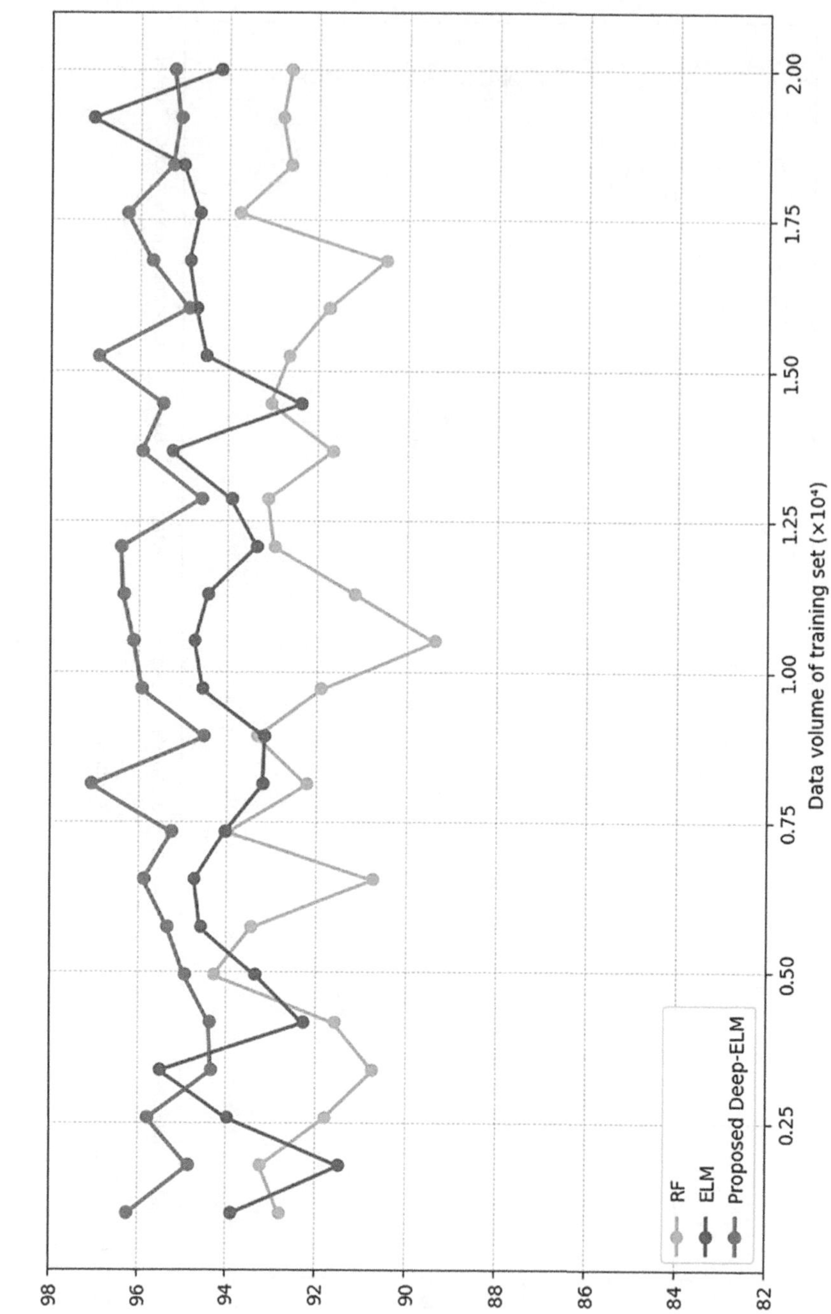

FIGURE 6.4 Accuracy comparison in % for different algorithms over varying volume of training set.

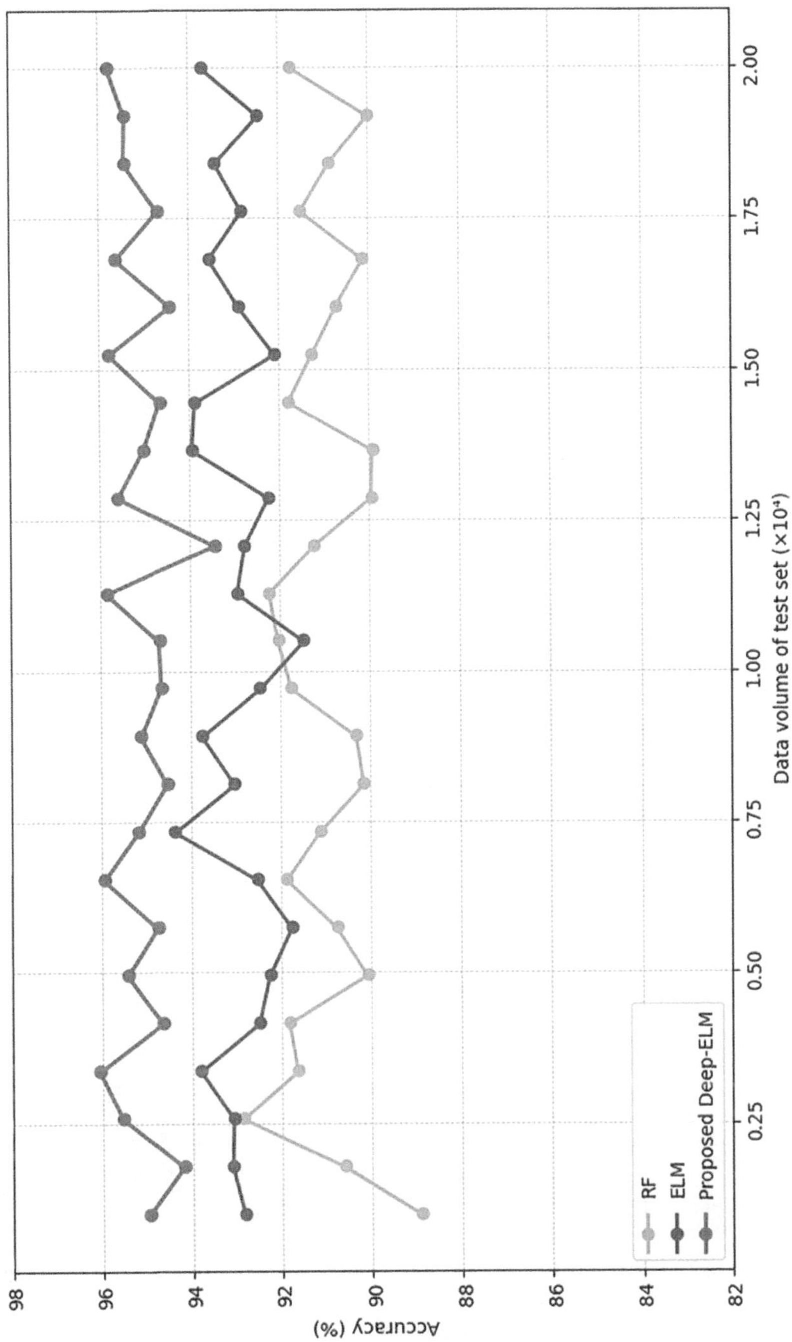

FIGURE 6.5 Accuracy comparison in % for different algorithms over varying volume of test set.

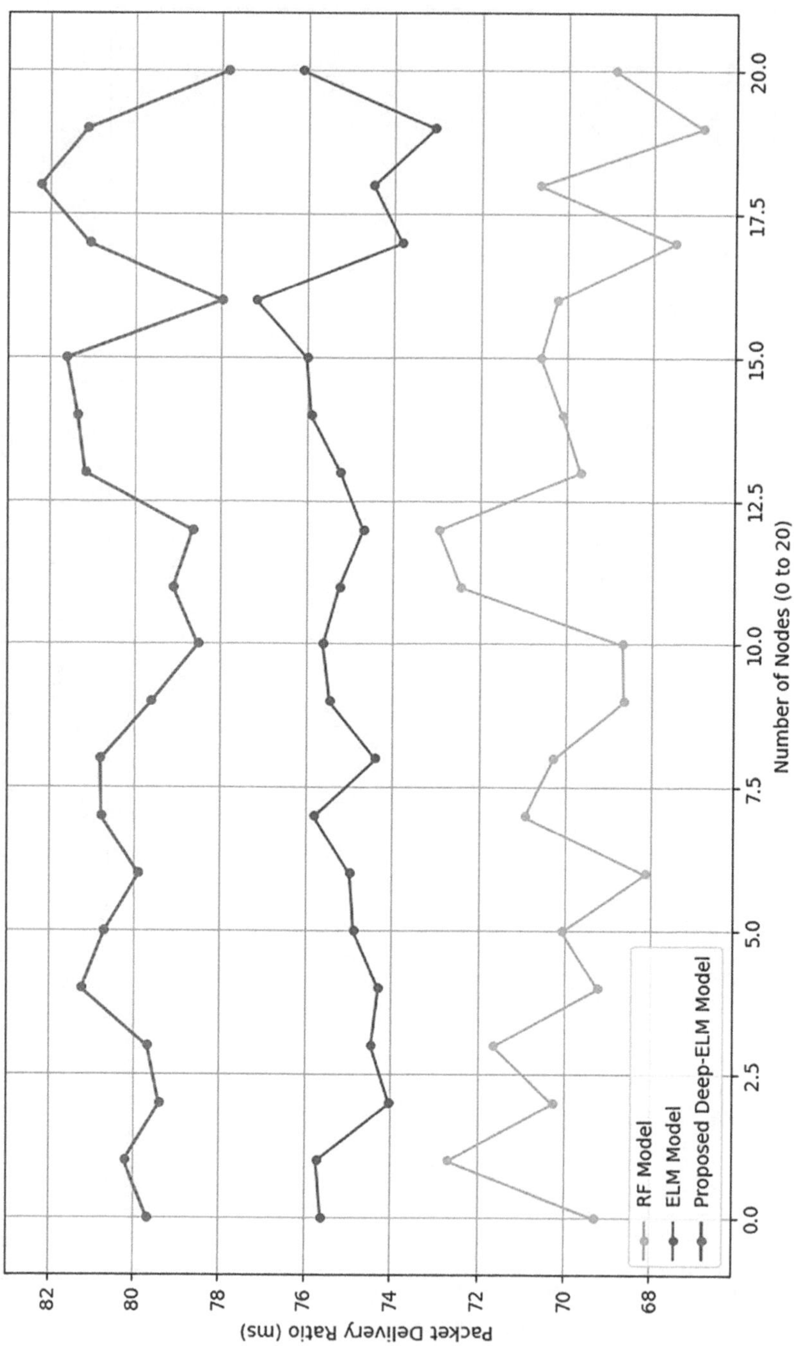

FIGURE 6.6 Comparison of packet delivery ratio over varying number of nodes.

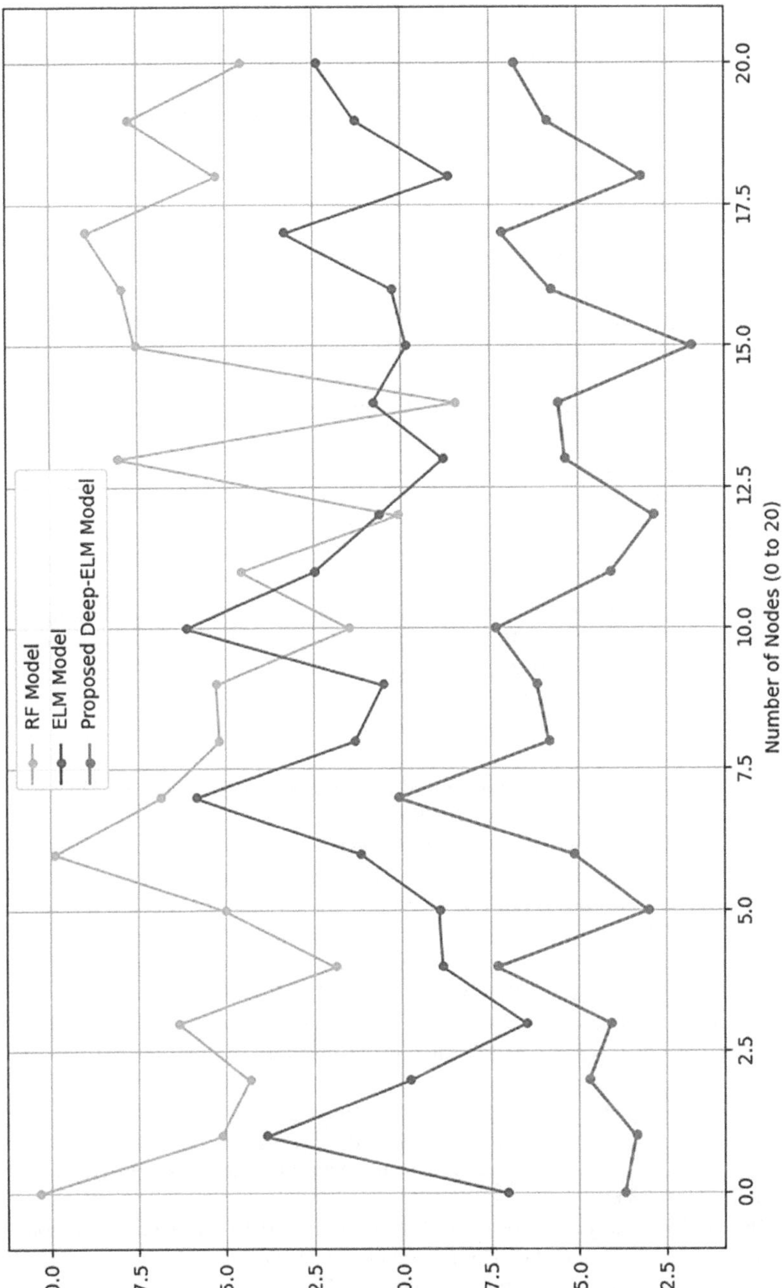

FIGURE 6.7 Comparison of packet delivery ratio over varying number of nodes.

REFERENCES

1. Zhang, Y., Li, P., & Wang, X. (2019). Intrusion detection for IoT based on improved genetic algorithm and deep belief network. *IEEE Access*, 7, 31711–31722.
2. Jha, S., Tariq, U., Joshi, G. P., & Solanki, V. K. (Eds.). (2022). *Industrial Internet of Things: Technologies, Design, and Applications*. CRC Press.
3. Kasongo, S. M. (2021). An advanced intrusion detection system for IIoT based on GA and tree based algorithms. *IEEE Access*, z9, 113199–113212.
4. Zhou, K., Liu, T., & Zhou, L. (2015). Industry 4.0: Towards future industrial opportunities and challenges. In *2015 12th International Conference on Fuzzy Systems and Knowledge Discovery*, FSKD, IEEE, pp. 2147–2152.
5. Jha, S., Ahmad, S., Abdeljaber, H. A. M., Othman, N. A., Nazeer, J., & Yadav, K. (2025). Enabling resilient wireless networks: OFDMA-based algorithm for enhanced survivability and privacy in 6G IoT environments. IEEE. https://ieeexplore.ieee.org/abstract/document/10445489
6. Aljuaid, W. H., & Alshamrani, S. S. (2024). A deep learning approach for intrusion detection systems in Cloud Computing Environments. *Applied Sciences*, 14(13), 5381.
7. Meddeb, A. (2016). Internet of Things standards: Who stands out from the crowd? *IEEE Communications Magazine*, 54(7), 40–47.
8. Koroniotis, N., Moustafa, N., & Sitnikova, E. (2020). A new network forensic framework based on deep learning for Internet of Things networks: A particle deep framework. *Future Generation Computer System*, 110, 91–106.
9. Shakya, S. R., & Jha, S. (2022). Challenges in Industrial Internet of Things (IIoT). In S. Jha, U. Tariq, G. P. Joshi, & V. K. Solanki (Eds.), *Industrial Internet of Things: Technologies, Design, and Applications* (pp. 21–40). CRC Press. https://doi.org/10.1201/9781003102267-2.
10. Dua, S., & Du, X. (2016). *Data Mining and Machine Learning in Cybersecurity*. CRC Press.
11. Kasongo, S. M. (2021). An advanced intrusion detection system for IIoT based on GA and tree-based algorithms. *IEEE Access*, 9, 113199–113212.
12. Hylving, L., & Schultze, U. (2013). Evolving the modular layered architecture in digital innovation: The case of the car's instrument cluster. In *ICIS*.
13. Shakya, S. R., & Jha, S. (2021). Internet of Things application in life sciences. In R. K. Tyagi, M. B. Patil, & T. K. Shah (Eds.), *The Role of Internet of Things (IoT) in Biomedical Engineering: Present Scenario and Future Scope* (Chapter 9). Academic Press. https://doi.org/10.1016/B978-0-12-817356-5.00012-7.
14. Hasan, M., Islam, M. M., Zarif, M. I. I., & Hashem, M. (2019). Attack and anomaly detection inIoT sensors in IoT sites using machine learning approaches. *Internet Things*, 7, 100059.
15. Mohammed, M., Khan, M. B., & Bashier, E. B. M. (2016). *Machine Learning: Algorithms and Applications*. CRC Press.
16. Agarap, A. F. (2018). Deep learning using rectified linear units (relu). arXiv preprint arXiv:1803.08375.
17. Jha, S., Nkenyereye, L., Joshi, G. P., & Yang, E. (2020). Mitigating and monitoring smart city using Internet of Things. *Computers, Materials & Continua,* 65(2), 1059–1079. www.techscience.com/cmc/v65n2/39862
18. Janarthanan, T., & Zargari, S. (2017). Feature selection in UNSW-NB15 and KDDCUP'99 datasets. In *Proc. IEEE 26th Int. Symp. Ind. Electron. (ISIE)*, pp. 1881–1886.

19. Arya, A., & Jha, S. (2022). An analytical review on data privacy and anonymity in "Internet of Things (IoT) enabled services." In 2022 6th International Conference on Computing, Communications and Security (ICCCS), IEEE, pp. 1–6. https://ieeexplore.ieee.org/document/9725770
20. Saxe, J., & Berlin, K. (2015). Deep neural network based malware detection using two dimensional binary program features. In *Malicious and Unwanted Software (MALWARE), 2015 10th International Conference on*, IEEE, 2015, pp. 11–20.
21. Shakya, S. R., & Jha, S. (2024). Introduction to Industrial Internet of Things (IIoT) – toward the future internet. In S. R. Shakya & S. Jha (Eds.), *Quality Assessment and Security in Industrial Internet of Things* (pp. 1–19). CRC Press. https://doi.org/10.1201/9781003530572
22. Zong, W., Chow, Y.-W., & Susilo, W. (2018). A two-stage classifier approach for network intrusion detection. In *Proc. Int. Conf. Inf. Secur. Pract. Exper.*, Springer, pp. 329–340.
23. Son, L. H., Jha, S., Kumar, R., Chatterjee, J. M., & Khari, M. (2019). Collaborative handshaking approaches between the internet of computing and the internet of things towards a smart world: A review from 2009–2017. *Telecommunication Systems,* 70(4), 617–634. https://doi.org/10.1007/s11235-018-0481-x
24. Poudyal, Y. R., Sharma, G., Adhikari, L. B., Chetry, A. B., & Jha, S. (2025). Developing Internet of Things (IoT) based smart government model for Nepal. *AIP Conference Proceedings,* 3224(1), 020007. https://doi.org/10.1063/5.0245896
25. Alanazi, M., & Aljuhani, A. (2022). Anomaly detection for Internet of Things cyberattacks. *Computers Materials & Continua*, 72(1), 261–279.

7 A Graph-based AI Generated Visual Manipulation Detector for UAV-based Industrial Systems

Manik Rakhra, Deepak Prashar, and Sudan Jha

7.1 INTRODUCTION

Over the last few years, the integration of Cognitive IoT sensor with Unmanned Aerial Vehicles (UAVs) [1, 2] has reached a high level of maturity and has already found many transformative applications in agriculture, logistics and infrastructure management. In particular, these sensors, that are powered by artificial intelligence (AI), artificial computing [3–5] and machine learning, give UAVs the ability to autonomously collect and process real time data from the ground up. This synergy has, thus, allowed UAVs to perform predictive maintenance, environmental monitoring, and precision agriculture with better accuracy and operational effectiveness [5, 6]. Also, the emergence of edge computing and the emergence of 5G networks provides UAVs with more edge local data processing capabilities [7], reducing latency and diminishes the reliance on cloud services. UAV deployments in industrial automation are becoming increasingly possible through these technological innovations that are furthering UAV deployment for the more adaptive, intelligent, and efficient operations [8].

Also, deep generative models have very fast progressed, generating better generated images, more realistic and diverse. Nonetheless, due to the inclusion of Cognitive IoT sensors in UAVs, it has been posited that AI generated images and videos might be used for exploitation in industrial settings [9]. A number of these threats include identity fraud, misinformation, political interference, as well as perturbing UAV collected data for logistics, industrial surveillance, and infrastructure inspection, namely, where goods or systems are dependent on being able to receive precise, real time visual inputs. Data in such environments could prove serious if manipulated and may lead to flawed decision making, inefficiencies in operations, and security risk to UAVs operating in the context of aerial computing systems [10].

To solve such problems, many algorithms to guarantee the credibility and maintain integrity of visual data [11, 12] and industrial IoT's anti-spoofing mechanisms [13, 14] have been developed. Most traditional detection techniques rely on extracting a

 DOI: 10.1201/9781003609773-7

global feature, at the cost of the intricacy of the relational feature between different data instance samples [15, 16]. Thus, these models perform well on the training datasets but suffer in generalization causing accuracy to drop sharply when exposed to previously seen images. In applications of industrial UAV, real time data integrity maintenance is vital, therefore, global and relational features must be captured in detection techniques for the robustness to be improved.

Current detection models have a low robustness and precision, especially when AI generated images have become more realistic and diverse. In this work, it is precisely an effort to propose a more general and versatile technique for spotting AI produced and AI manipulated images [17, 18] by training neural networks using fake images generated by a single generative model. Yet, it is known that each of the AI models used for the synthetic generation generates traces that are difficult to detect. With these subtle discrepancies, these peculiarities in the color, texture, or other image details, they have to be detected through advanced analytical methods. Previous studies [19, 20] in this area have primarily focused on the image artifacts. Current detection systems nevertheless exhibit challenges in their utilization in real world UAV applications for industrial operations. Real world environments bring much more variability and complexity than controlled conditions such that many existing models were developed under those conditions that are limited and unprocessed [21–23].

GANs and diffusion models generate extremely realistic images and videos with fine quality and localized artifacts, that have become much more complex to synthesize and hard to detect. Moreover, detection models for UAV based industrial systems also need to be robust against various image modifications, adversarial attacks as well as environmental fluctuations. Such unpredictable conditions, alongside high stakes applications where unreliable or inconsistent performance severely impacts results, ensure that consistent and reliable performance is highly important. With this, the proposed dual stream architecture makes the gap between theoretical progress and industrial deployment across these key challenges, which provide a scalable and robust dual stream solution for industrial applications [24–26].

In this paper, we propose a novel dual stream framework which attempts to provide a means of improving both feature representation as well as diversity in cognitive sensors deployed in UAV systems. The integration of the CLIP's robust feature extraction skills with the GCN's relationship learning potential helps improve detection accuracy. We train the detector using images produced by ProGAN and assesses them on various generative models: GANs, diffusion models, DeepFakes. The proposed method is designed with strong robustness to image distortions and good performance in noisy environments. Pre-training models like CLIP is incorporated to maintain scalability and allow for deployment within unmanned aircraft systems having constricted computational computing resources. Through the correct identification of AI generated visual content, our framework protects data authenticity for critical decisions in the industrial, government, public sector. Moreover, the adaptability across heterogeneous operational conditions including Gaussian blurring, JPEG compression, Gaussian noise and cropping are robustness evaluated.

This research has the following principal contributions:

We propose a dual stream framework that makes use of CLIP for the global feature extraction and learns relational feature with GCN, extending the capabilities of future UAV cognitive sensor systems for next generation industrial purpose. Specifically, we introduce a graph based structure to learn the latent features for data samples which are shared with same label and dissimilar from each other, further connecting GCN with the task of graph based metric optimization, for improving the detecting performance in complicated environments. We provide comprehensive experimental analyses on several datasets, proving the effectiveness of the design of our approach to detection of AI generated images with high accuracy, using both global and relational feature representation.

7.2 RELATED WORK

we develop previous advances in generative detection techniques in terms of the graph neural networks (GNNs) and metric learning, which have come to solve increasingly sophisticated detection scenarios.

7.2.1 Cognitive Sensors for UAVs

Research interest has arisen in the advancement of cognitive sensors for UAV technology because they enable enhancing the autonomous decision making in the complex operational environment. Powered by AI and machine learning, these sensors help the UAVs interpret real time data, have autonomy to adapt to the dynamic surroundings and performed tasks with improved autonomy. Implementation of these is spread across several fields of application, from precision agriculture to disaster relief to industry surveillance. Cognitive sensors can be integrated into the UAVs that allow them to identify anomalies, support predictive maintenance, improve energy efficiency, and overall make operations more streamlined through reduced need for human oversight.

Studies abound as to the benefits of cognitive sensors in improving the performance of UAVs. These sensors for instance have been used to evaluate the status of crops by the factors depending upon soil moisture content, temperature variations and aerial imaging data [27]. Moreover, the YOLO algorithm has also been utilized to do object detection in UAV based systems [21, 28]. However, there are still some obstacles. A major issue is the data processing in highly unpredictable and changing environment [29]. Despite the demonstrated analytical potential of AI driven cognitive sensors in a controlled setting, their effectiveness in real world applications under conditions of weather, signal interference, and dimensional variations of the environment. In low visibility condition or areas with electromagnetic interference, Yan et al. found that UAV cognitive systems are troublesome to keep the collected data accurate [30].

The UAV systems also have significant security and privacy risks involved. Cognitive UAVs are dependent on wireless networks and cloud based data processing and hence they are subjected to cyber threats such as those that can compromise data authenticity and system stability [31]. Additionally, visual content of AI generated images and videos adds an additional level of challenge due to the possible deception of cognitive sensing, and hence also flawed decision making in mission critical operations including assessment of critical structures and environmental monitoring. The UAV systems are also prone to such adversarial attacks that render it very difficult to develop cognitive sensors that are able to distinguish between genuine and artificial content, thus undermining the reliability and accuracy of UAV functions [18, 20].

7.2.2 Identification of AI-Generated Images

Generative Adversarial Networks (GANs) and diffusion models recently made a big leap in realism of generated images, yet the advancements make the task of the cognitive sensors in UAVs a more difficult one. However, while AI generated images look very realistic, these images have artifacts and residual patterns among them that can be used as indicators of detection. This gives rise to the development of different methodologies that separate synthetic images from real ones in three major categories:

The fingerprint based techniques [32, 22] Each generative model generates a different digital trace that you can check for detection purposes. FingerprintNet [23] was also introduced by Jeong et al. to use frequency spectrum analysis to improve on accuracy for identifying synthetic content. Likewise, Sinitsa et al. [32] create Deep Image Fingerprint (DIF), a technique for finding artifacts unique to convolutional neural network (CNN) and thus able to identify images produced from the same generative model. Wang et al. also [24] investigated some of the fundamental characteristics of the diffusion models to develop a method for their detection.

Spatial Based Techniques [25, 26], These methods focus on the spatial attributes like the texture patterns and gradient difference. In high-texture and low-texture regions of a synthetic image, Zhong et al. [25] evaluated pixel correlations and looked for inconsistencies in synthetic images. Similarly, as shown by Nguyen et al. [26] feature extraction using gradient based approach can be used to distinguish an image generated from different sources by using deep learning. In another [33]', a deep learning based framework that recognizes between natural images and these artificial generated pictures, was proposed. Additionally, an explainable detection method based on a CNN also works better in detecting synthetic images as reported in [34].

Frequency Based Techniques [35, 36] convert the images into the frequency domain, so as to extract the hidden information, which may not be easily accessible in the spatial domain. In their paper, Jeong et al. [35] presented the Bilateral High Pass Filter (BiHPF) used to improve frequency-based artifacts and increase detection accuracy. Like Bammey et al. [36], we applied cross difference filters to these artifacts as well and so the images created by AI stand out more.

Merging these detection methods to cognitive sensors could greatly extend UAV security where it would be capable of detecting and mitigating threats brought on by AI image generation. This guarantees that the critical data used in UAV operations is not subject to manipulation and error in decision making.

Finally, we evaluate the proposed method with images generated by diffusion models to get a comprehensive analysis of efficacy. In particular, we choose seven generative models from GenImage dataset: ADM [49], Glide [50], Midjourney [51], SDV1.4 [52], SDV1.5 [52], VQDM [53], and Wukong [54], to evaluate how accurately the model detects these generative models. The publicly available FF++ dataset from [48] is used as a source for deepfake samples, and images produced by Midjourney and Wukong are acquired from several online sources.

7.4.2 Comparison with State-of-the-arts

Evaluation Metric. We also adopt the evaluation criteria used by previous studies [40, 55] to determine the authenticity of generated images. Key metrics are employed specifically to evaluate the effectiveness of our method, namely, classification accuracy (ACC) and average precision (AP). The results presented also show that the test set is balanced accordingly, i.e. there is an equal number of real and synthetic images per generative model.

Implementation Details. As a backbone network, we use the CLIP:ViT-B/16 model pre-trained. All images are preprocessed to 224 × 224 pixels. Random augmentations such as Gaussian blur, JPEG compression, and horizontal flip are applied on random basis with probability set at training time. To make the image inputs standard, we first apply center cropping for testing.

The Adam optimizer is used to optimize the model with learning rate initialized to 0.0001 and dynamically trained. The batch size is set to 128. There are three sane parameters of our two layer GCN module, each has input channels, hidden channels and output channels. The feature dimensions extracted by CLIP/16 correspond to the input channels; the number of hidden channels in the hidden layer is set to be 512. The number of channels of the output layer is set up to 50.

To test all of this, experiments are run on an Ubuntu 18.04 machine using PyTorch. These include two dedicated servers with two NVIDIA RTX 3090 Ti GPUs and two NVIDIA RTX 4090 GPUs for computational resources which make us effective with their processing and scalability for our simulations.

Evaluation Metric. We also adopt the evaluation criteria used by previous studies [40, 55] to determine the authenticity of generated images. Key metrics are employed specifically to evaluate the effectiveness of our method, namely, classification accuracy (ACC) and average precision (AP). The results presented also show that the test set is balanced accordingly, i.e. there is an equal number of real and synthetic images per generative model.

Implementation Details. As a backbone network, we use the CLIP:ViT-B/16 model pre-trained. All images are preprocessed to 224 × 224 pixels. Random augmentations such as Gaussian blur, JPEG compression, and horizontal flip are applied on random basis with probability set at training time. To make the image inputs standard, we first apply center cropping for testing.

The Adam optimizer is used to optimize the model with learning rate initialized to 0.0001 and dynamically trained. The batch size is set to 128. There are three sane parameters of our two layer GCN module, each has input channels, hidden channels and output channels. The feature dimensions extracted by CLIP/16 correspond to the input channels; the number of hidden channels in the hidden layer is set to be 512. The number of channels of the output layer is set up to 50.

To test all of this, experiments are run on an Ubuntu 18.04 machine using PyTorch. These include two dedicated servers with two NVIDIA RTX 3090 Ti GPUs and two NVIDIA RTX 4090 GPUs for computational resources which make us effective with their processing and scalability for our simulations.

7.4.3 Ablation Study

Impact of Individual Modules, It examines the role of each of the components of the proposed framework through ablation studies, to quantify the effectiveness of it. The results of these experiments are presented in Table III. We built a baseline classifier that only used the extracted features for feature extraction and used the pretrained CLIP/16 model. While the baseline from this is strong in detecting GAN generated images, it was severely limited when exposed to previously unseen diffusion models. By fusing CLIP model with GCN module, we observed drastic increases of 8.9 percent in classification accuracy (ACC) and 6.2 percent in average precision(AP). These improvements demonstrate the effectiveness of graph convolution module in refining and learning feature representations from the CLIP output, so that it can better comprehend structural relationships among data samples.

At the final evaluation stage, a graph topology based metric learning approach has been included in the model so as to automatically prioritize samples that are hard to classify. The performance of this optimization improved significantly with 99.1% of classification accuracy and 99.7% of average precision. Such outcomes confirm our proposal's superb detection performance on distinguishing real and AI-generated images. To better understand if our graph topology based metric learning method could have a contribution compared with conventional loss functions, we evaluated it based on loss functions. To replace Lmea, we kept the same network architecture and training procedures yet replace Lmea with Contrastive Loss and Triplet Loss. Table IV is the result of such tests. The detection accuracy of proposed metric loss function achieves a great average of 97.7% on specific databases. Despite this, Contrastive Loss and Triplet Loss proved to be successful on detecting GAN generated images, but were not good at detecting images of other generative models. Our findings show that our custom loss function is able to capture the complex relationships and structural dependencies among the graph topology nodes, as a result of which it boosts the model's performance to detect AI generated images in hard settings.

7.4.4 Robustness to Post-Processing Operations

Most of the time, images are processed with a bunch of post processing operations. We further found the robustness of our proposed method to the unforeseen changes by applying four common perturbations to the test images: Gaussian blur (σ=1), JPEG compression (quality=50), Gaussian noise (variance=10), and image cropping (r=0.5) Table V summarizes the model performance under perturbations. The model is very robust to blurring and noise based distortion as indicated by the findings. Although a small drop in ACC (7.2%) and AP (5.2%) was observed over JPEG compression and image cropping, the model significantly outperformed state of the art models in in accordance with the number of training images used for training, as shown on Figure. Such variations of the testing distribution only greatly reduced robustness and adaptability of our model, while stable detection performance on various post processing transformations.

7.4.5 Qualitative Analysis

We used t-distributed Stochastic Neighbor Embedding (t-SNE) to better understand the quality of our suggested method, where we show the visualization of the distribution of feature vectors in the feature space. In comparison to the baseline of only using CLIP/16, the illustration was carried out. This experiment was based on randomly selecting 1,000 samples of test dataset for GauGAN.

It is shown that our model differentiates better between real and AI generated images than the baseline. Our approach in turn establishes a nice well defined decision boundary for its enhanced differentiation, which indeed reflects strong strength in detection of fake content. It is attributed to the better feature representation learning in graph convolution module, which further enriches the embedding space by more unique and discriminative set of embeddings. In the t- SNE visualization, generated and real images have a more structured and decisive spatial arrangement with more comfortable separation of the two.

7.4.6 Privacy and Security

Although UAVs and Cognitive Internet of Things (CIoT) environments are highly beneficial in their convergence in at least industrial applications, they also raise great concerns in terms of privacy and security. Such identification of AI tampered visual information in UAVs will improve the integrity and credibility of the information that UAVs gather, reducing the risk of deceptive data manipulation which can impact decision making in sensitive environments. However, installation of such detection mechanisms requires licensing to strict privacy laws, mainly in use cases where UAVs operate in a surrounding the entails personal or confidential information. Additionally, the detection model must be protected from adversarial attacks which could make the AI generate content incorrectly, be used to exploit vulnerabilities of the detection model. Future versions of the proposed framework can use federated learning to promote data privacy without decreasing robustness to old attacks, as well as using adversarial training methods to counter the possible future threats. Dealing with these privacy and security issues will ensure that IIoT systems are made more reliable and resilient. We find that besides being slower on training time, the new method is more accurate but at the expense of more expensive computational complexity. However, it is not easy to make direct comparisons since there are differences in parameter settings for each of the models, which heavily affects training time as a whole. Furthermore, experiments on two types of server environments were run, and direct comparisons were even more difficult. While on the same hardware conditions, most established models end training in less than a day, our method is slower. However, once trained, our testing efficiency is comparable to all models. We tested all methods almost real time within our environment.

7.5 CONCLUSION

This paper introduces a dual stream framework, which is created by combining Graph Convolution Networks (GCN) with Contrastive Language-Image Pretraining (CLIP) for enhancing detection efficiency. With this method, global visual features and relational dependencies can be effectively excerpted semantically with CLIP, and reason relationally with context in mind through graph-based architectures. Various contributions reformulate the presented problem and propose our solution, which considerably improves the identification of AI generated images, thereby improving the security and reliability of Unmanned Aerial Vehicle (UAV) systems in IIoT applications. Extensive experimental tests show that the proposed model does generalize well to what hasn't been seen, e.g. DeepFakes, and it is robust to these real world perturbations. It is compared to state of the art baseline models and shows superiority. Once integrated into UAV cognitive sensors for aerial computing, data processing and decision making can be simplified, enabling the IIoT Application development in complex operational environment.

Our model applications to a wide variety of industrial applications is not only suited to critical infrastructure monitoring as well as autonomous inspection and remote sensing missions. To make such decisions, it is important to be able to detect AI manipulated content in order to get the highest accuracy and reduce the security risks with flavoured data streams. Graph based relational learning used in the application of the model makes the model interpretable and enhances detection accuracy due to the fact that it models complex dependence among the components of an image better than traditional features extraction approaches. Moreover, the framework is also scalable by design and capable of large-scale autonomous UAV networks for real time IoT settings.

Although there are these strengths, however, there are also problems. The method is highly computationally intensive and having it run on UAVs with limited resource is extremely impractical. More optimization may be required (e.g. using more parameters such as at the cost of worse computational complexity to better represent the topological structure). Future work needs to tackle these limitations with hardware accelerations, such as edge AI processors, FPGA based implementation, and distributed cloud edge architecture to conduct real time application on resource limited UAV platform. However, the present method reaches an appropriate compromise between efficiency and computational feasibility under such constraints. However, using the limits to the advancement of cloud computing, 6G networks, edge AI technologies, these limitations should be removed and allow seamless compatibility with next generation UAV systems. Additionally, it can improve privacy preserving distributed UAV collaborative learning by further applying federated learning methodologies toward federating distributed UAV networks for secure model updates without divulging the sensitive operational details.

Future work will remain in utilizing the advanced generative models to produce images that are used to improve detection accuracy and robustness. Further improvements in detection models will be necessary to fight against the manipulation techniques that are constantly developing in deep generative architectures such those like Diffusion Models, GAN variations and Transformer based image synthesizers. With adaptive learning mechanisms such as self supervised learning and meta learning the detection framework will leverage for adaptive representations of feature evaluated in real time. Further, a broader examination of the data scope by bringing in multimodal data (textual, audio, thermal imaging inputs) will result in the development of an end-to-end solution to detect the synthetic content in heterogeneous IoT data streams. Overall, the proposed dual stream architecture is a substantial development in AI generated content detection for UAV based in IIoT networks. This research leverages emerging AI based techniques and extends detection to a multimodal paradigm to address the computation challenges in order to lead to the next generation of secure intelligent UAV surveillance system for high risk industrial environment.

REFERENCES

1. S. Policepatil, J. Sharma, B. Kumar, D. Singh, S. Pramanik, A. Gupta and B. S Mahabub, "Financial Sector Hyper-Automation: Transforming Banking and Investing Procedures," *Examining Global Regulations During the Rise of Fintech*, IGI Global, 2025, pp. 299–318.
2. A. Wahab, "Impact of Artificial Intelligence on Indian Banking Sector-A Study of Banks," *International Research Journal on Advanced Engineering and Management (IRJAEM)*, vol. 2, no. 05, pp. 1261–1268, 2024.
3. P. Mehta and A. K. Jha, "The Future of Finance: Exploring The Role of AI and Automation in Revolutionizing Indian Banking Processes," *Educational Administration: Theory and Practice*, vol. 30, no. 2, pp. 492–499, 2024.
4. K. R. P. Vittala S.Ahmad, R. Seranmadevi and A. M. Tyagi, "Emerging Technology Adoption and Applications for Modern Society Towards Providing Smart Banking Solutions," *Enhancing Medical Imaging with Emerging Technologies*, IGI Global, 2024, pp. 315–329.
5. S. K. Singh, J. K. Parida and S. Shekhar, "The Landscape of AI in Indian Banking Sector: A Theoretical Perspective," *Journal of Informatics Education and Research*, vol. 4, no. 1, pp. 1091–1099, 2024.
6. N. L. Rane, S. P. Choudhary and J. Rane, "Artificial Intelligence-Driven Corporate Finance: Enhancing Efficiency and Decision-Making Through Machine Learning, Natural Language Processing, and Robotic Process Automation in Corporate Governance and Sustainability," *Studies in Economics and Business Relations*, vol. 5, no. 2, pp. 1–22, 2024.
7. R. Kumari, K. Sharma and R. Kumar, "Social and Ethical Implications of AI in Finance for Sustainability: Convergence of Culture in the Banking Sector After Adaptation of Artificial Intelligence," *Social and Ethical Implications of AI in Finance for Sustainability*, IGI Global, 2024, pp. 141–155.
8. D. Singh, A. Singh, M. Rakhra, G. S. Cheema and A. Khamparia, "Predictions on the Future of Agriculture and Recent Developments in Agricultural Technology," *International Conference on Artificial Intelligence of Things For Smart Societies*. Cham: Springer Nature Switzerland, 2024.
9. V. Kanaparthi, "Transformational Application of Artificial Intelligence and Machine Learning in Financial Technologies and Financial Services: A Bibliometric Review," *arXiv preprint arXiv:2401.15710*, 2024.

10. L. D. Oyeniyi, C. E. Ugochukwu and N. Z. Mhlongo, "Transforming Financial Planning with AI-Driven Analysis: A Review and Application Insights," *Finance & Accounting Research Journal*, vol. 6, no. 4, pp. 626–647, 2024.
11. T. Sarkar, R. Kumar, M. S. Kumar, S. Aggarwal, A. Sandhya and A. M. Shukla, "Review Paper of Performance Analysis in Wireless Sensor Networks," *Kilby*, vol. 100, p. 7th, 2023.
12. D. Singh et al., "of Agriculture and Recent Developments in Agricultural," *The Smart IoT Blueprint: Engineering a Connected Future: Guiding Principles and Practical Strategies for Seamless Integration 297*, 2024.
13. S. Jha, S. Routray, H. A. Abdeljaber and S. Ahmad, "A Novel Approach for Decision Support System in Cricket Using Machine Learning," *International Journal of Computer Applications in Technology*, vol. 69, no. 1, pp. 86–92, 2022.
14. H. Kaur, T. Sarkar, B. Kaur, A. Singh, G. S. Cheema and S. Das, "Using Machine Learning Techniques in Business Analytics: Lessons from Agile Project Management," *2024 7th International Conference on Contemporary Computing and Informatics (IC3I)*, Greater Noida, India, 2024, pp. 214–220, doi: 10.1109/IC3I61595.2024.10828636.
15. T. Sarkar, M. Rakhra, V. Sharma, A. Singh, K. Jairath and A. Maan, "Comparing Traditional vs Agile Methods for Software Development Projects: A Case Study," *2024 7th International Conference on Contemporary Computing and Informatics (IC3I)*, Greater Noida, India, 2024, pp. 221–227, doi: 10.1109/IC3I61595.2024.10829321.
16. S. Takkar, A. S. Rayat, V. Sharma, T. Sarkar, K. Jairath and A. Singh, "Problems and Optimal Methodologies Agile Project Management Solution in Publicly accessible Sector: A Comprehensive Evaluation of the Literature," *2024 7th International Conference on Contemporary Computing and Informatics (IC3I)*, Greater Noida, India, 2024, pp. 207–213, doi: 10.1109/IC3I61595.2024.10828832.
17. S. Singh, T. Sarkar, M. Mangla, M. Rakhra, A. Singh and K. Jairath, "FinTech Edge: Utility Computing & Artificial Intelligence Technologies for Smart Financial Acquisition & Blockchain in the Financial Industries," *2024 7th International Conference on Contemporary Computing and Informatics (IC3I)*, Greater Noida, India, 2024, pp. 228–234, doi: 10.1109/IC3I61595.2024.10828908.
18. S. Jha et al., "Enabling Resilient Wireless Networks: OFDMA-Based Algorithm for Enhanced Survivability and Privacy in 6G IoT Environments," *IEEE Transactions on Consumer Electronics*, vol. 70, no. 1, pp. 3810–3819, Feb. 2024, doi: 10.1109/TCE.2024.3370414.
19. V. K. Singh, B. Moharana, T. Sarkar, A. Maan, A. Bhattacherjee and M. Rakhra, "Practical Assessment of Large Language Models in Cloud Computing Using Real-World Data Applications," *2024 International Conference on Cybernation and Computation (CYBERCOM)*, Dehradun, India, 2024, pp. 356–362, doi: 10.1109/CYBERCOM63683.2024.10803266.
20. S. Talwani, M. Rakhra and T. Sarkar, "Perspectives on the Future of Agriculture and Recent Developments in Agricultural Technology," *2024 11th International Conference on Reliability, Infocom Technologies and Optimization (Trends and Future Directions) (ICRITO)*, Noida, India, 2024, pp. 1–6, doi: 10.1109/ICRITO61523.2024.10522413.
21. A. K. Verma, M. Rakhra, A. Bhattacherjee, A. Maan, T. Sarkar and V. K. Pandey, "A Suggested Model for Using Multi Factor Authentication Framework in Cloud Computing for SME," *2024 International Conference on Cybernation and Computation (CYBERCOM)*, Dehradun, India, 2024, pp. 192–197, doi: 10.1109/CYBERCOM63683.2024.10803146.

22. A. K. Verma, T. Sarkar, M. Rakhra, V. K. Pandey and P. Chakrabarty, "Studying the Impact of Voice-Activated Chatbot's with Artificial Intelligence on Consumer Interaction in India," 2024 4th International Conference on Technological Advancements in Computational Sciences (ICTACS), Tashkent, Uzbekistan, 2024, pp. 1708–1713, doi: 10.1109/ICTACS62700.2024.10841074.
23. B. Moharana, V. K. Singh, T. Sarkar, A. Hassan and V. K. Pandey, "Predicting Indian Master's and PhD Enrollment using Data Mining & Machine Learning Approaches," 2024 4th International Conference on Technological Advancements in Computational Sciences (ICTACS), Tashkent, Uzbekistan, 2024, pp. 1714–1720, doi: 10.1109/ICTACS62700.2024.10841260.
24. T. Sabharwal, R. Gupta, L. H. Son, R. Kumar and S. Jha, "Recognition of Surgically Altered Face Images: An Empirical Analysis on Recent Advances," *Artificial Intelligence Review*, vol. 52, no. 2, pp. 1009–1040, Aug. 2019, doi: 10.1007/s10462-018-9660-0.
25. H. C. Bhandari, Y. R. Pandeya, K. Jha, S. Jha and S. Ahmad, "Exploring Non-Euclidean Approaches: A Comprehensive Survey on Graph-Based Techniques for EEG Signal Analysis," *Journal of Advances in Information Technology*, vol. 15, no. 10, pp. 1089–1105, 2024.
26. S. Jha, S. Routray and S. Ahmad, "An Expert System-Based IoT System for Minimisation of Air Pollution in Developing Countries," *International Journal of Computer Applications in Technology*, vol. 68, no. 3, pp. 277–285, 2022.
27. B. Moharana, V. K. Singh, T. Sarkar, A. Maan, M. Rakhra and P. Chakrabarty, "A Comprehensive Review on Performance Analysis of Enterprise Cloud Computing Architecture," *2024 International Conference on Cybernation and Computation (CYBERCOM)*, Dehradun, India, 2024, pp. 184–191, doi: 10.1109/CYBERCOM63683.2024.10803178.
28. V. Sharma, M. Rakhra, V. Sharma, T. Sarkar, A. Singh and K. Jairath, "Hybrid Approaches for Stocks Price Prediction: A New Institutive Way of Neural Architecture Design," *2024 International Conference on Electrical Electronics and Computing Technologies (ICEECT)*, Greater Noida, India, 2024, pp. 1–6, doi: 10.1109/ICEECT61758.2024.10739249.
29. M. Rakhra, T. Sarkar, V. Sharma, P. K. Angra, A. Kumar and P. Singh, "An In-Depth Analysis of Local Mechanic Support For Immediately Apparent Repairs And Scheduled Maintenance Services," *2024 International Conference on Electrical Electronics and Computing Technologies (ICEECT)*, Greater Noida, India, 2024, pp. 1–7, doi: 10.1109/ICEECT61758.2024.10739024.
30. S. Jha, N. Jha, D. Prashar, S. Ahmad, B. Alouffi and A. Alharbi, "Integrated IoT-Based Secure and Efficient Key Management Framework Using Hashgraphs for Autonomous Vehicles to Ensure Road Safety," *Sensors*, vol. 22, p. 2529, 2022, www.mdpi.com/1424-8220/22/7/2529
31. S. Jha, "Chapter 1 – Introduction to Smart Soft Computing in Engineering: Embracing Intelligence and Efficiency," *Soft Computing in Smart Manufacturing and Materials*, S. Jha, S. Datta, D. Prashar, S. R. Shakya, and V. E. Balas, Eds. Elsevier, 2025, pp. 1–10, doi: 10.1016/B978-0-443-29927-8.00001-1.
32. S. Jha, D. Prashar, H. V. Long and D. Taniar, "Recurrent Neural Network for Detecting Malware," *Computers & Security*, vol. 99, p. 102037, Dec. 2020, doi: 10.1016/j.cose.2020.102037.
33. S. Jha, "Introductory Chapter: Federated Earning – Bridging the Gap between Privacy and Collaboration in Machine Learning," *Federated Learning - A Systematic Review*, 2025. IntechOpen. Available at: www.intechopen.com/chapters/1221396

34. S. Jha, "Introduction to Smart Soft Computing in Engineering: Embracing Intelligence and Efficiency," *Soft Computing in Smart Manufacturing and Materials*, Elsevier, 2025, pp. 1–10, doi: 10.1016/B978-0-443-29927-8.00001-1
35. S. Jha, et al. "Enabling Resilient Wireless Networks: OFDMA-Based Algorithm for Enhanced Survivability and Privacy in 6G IoT Environments," *IEEE Transactions on Consumer Electronics*, 2024, doi: 10.1109/TCE.2024.3370414
36. S. Jha, et al. "Ensemble Learning-Based Hybrid Segmentation of Mammographic Images for Breast Cancer Risk Prediction using Fuzzy C-Means and CNN Model," *Journal of Healthcare Engineering*, 2023. doi: 10.1155/2023/1491955
37. S. Jha, "Synthesis, Characterization, and Performance of Chitosan/Nylon 6/ Polyurethane Blend for the Removal Of Chromium (VI) and Lead (II) Ions from Aqueous Solutions for Enhanced Kinetic Adsorption Studies," *Water Reuse*, 2023, doi: 10.2166/wrd.2023.019
38. S. Jha, et al. "Integrated IoT-Based Secure and Efficient Key Management Framework Using Hashgraphs for Autonomous Vehicles to Ensure Road Safety," *Sensors*, vol. 22, no. 7, p. 2529, 2022, doi: 10.3390/s22072529
39. S. Jha, et al. "A Post Covid Machine Learning Approach in Teaching and Learning Methodology to Alleviate Drawbacks of the E-Whiteboards," *Journal of Applied Science and Engineering*, vol. 25, no. 2, pp. 285–294, 2022, doi: 10.32604/ cmc.2022.024970
40. S. Jha, et al, "Real Time Object Detection and Tracking System for Video Surveillance System," *Multimedia Tools and Applications*, 2021, doi: 10.1007/ s11042-020-09749-x
41. S. Jha, et al. "Neutrosophic Approach for Enhancing Quality Of Signals," *Multimedia Tools and Applications*, 2019, doi: 10.1007/s11042-019-7375-0
42. S. Jha, et al. "Neutrosophic Image Segmentation with Dice Coefficients," *Measurement*, vol. 134, pp. 762–772, 2019, doi: 10.1016/j.measurement.2018.11.006
43. S. Jha, et al. "Deep Learning Approach for Software Maintainability Metrics Prediction," *IEEE Access*, vol. 7, pp. 61840–61855, 2019, doi: 10.1109/ ACCESS.2019.2913349
44. S. Turpati, B. Geetha Rani, A. V. Prabu, A. Mukherjee, S. Jha & K. C. T. Swamy, "Review of 6G Wireless Communication System With Artificial Intelligence," *Internet Technology Letters,* vol. 8, no. 6, e70127, 2025, doi: 10.1002/itl2.70127
45. T. Kaur, R. Garg, D. Prashar, S. Jha, S. Ahmad & J. Nazeer, "Healthcare-as-a-Service Provisioning using Cloud-of-Things: A Contemporary Review of Existing Frameworks based on Tools, Services and Diseases," *Scalable Computing: Practice and Experience*, 2025, doi: 10.12694/scpe.v26i5.4823
46. A. R Paudel, P. Chalise, R. Poudyal and S. Jha, "A Proposed Algorithm for Object Detection System for Enhanced Mobility of Visually Impaired Individuals—'Aadarsha Dristi'," *Innovative Computing and Communications (ICICC 2025)*, A.E. Hassanien, S. Anand, A. Jaiswal, & Prabhat Kumar Eds. Springer, 2025, pp. 1–12, doi: 10.1007/978-981-96-6883-0_44
47. S. Ahmad, M. Alharbi, S. Jha, & A. Ali, "*Federated Learning*," IntechOpen. 2025, doi: 10.5772/intechopen.1003284
48. S. Ghimire, N. Bhurtel, S. Jha, S. Ahmad, H. A. M. Abdeljaber and J. Nazzer, "Minimizing Energy Consumption in Fixed Node Networks Using a Novel Neutrosophic Model: Comparative Analysis with Standard and Existing Algorithms," *International Journal of Information Technology*, 2025, https://link.springer.com/article/10.1007/s41870-024-02368-x

49. H. C Bhandari, Y. R Pandeya, K. Jha and S. Jha, "Recent Advances in Electrical Engineering: Exploring Graph Neural Networks for Weather Prediction in Data-Scarce Environments," *Environmental Research Communications*, 2024, doi: 10.1088/2515-7620/ad8108
50. Bhandari, H. C., Pandeya, Y. R., Jha, K., Jha, S., & Ahmad, S. (2024). Exploring Non-Euclidean Approaches: A Comprehensive Survey on Graph-Based Techniques for EEG Signal Analysis. *Journal of Advances in Information Technology,* vol. 15, no. 10, pp. 1089–1105. doi: 10.12720/jait.15.10.1089-1105
51. Sharma, A., Chopra, S. R., Sapate, S. G., Arora, K., Rahmani, M. K. I., Jha, S., ... & Nazeer, J. (2024). Artificial Intelligence Techniques for Landslides Prediction using Satellite Imagery. *IEEE Access*. doi: 10.1109/ACCESS.2024.3446037
52. Dhande, A. P., Malik, R., Saini, D., Garg, R., Jha, S., Nazeer, J., & Ahmad, S. (2024). Design of a High-Efficiency Temporal Engine for Real-Time Spatial Satellite Image Classification Using Augmented Incremental Transfer Learning for Crop Analysis. *SN Computer Science,* vol. 5, no. 5, p. 585. doi: 10.1007/s42979-024-02939-6
53. Turpati, S., Roy, A., Swamy, K. C. T., Rahmani, M. K. I., Rehman, M. U., Jha, S., ... & Aljohani, Y. E. (2024). Robust Non-Linear Active Function-Based FxRLS for Impulsive Active Noise Control. *IEEE Access,* vol. 12, pp. 43222-43234. doi: 10.1109/ACCESS.2024.3378067
54. Rajagopal, A., Ahmad, S., Jha, S., Abdeljaber, H. A. M., & Nazer, J. (2023). AI-Based Secure Analytics of Clinical Data in Cloud Environment: Towards Smart Cities and Healthcare. *Journal of Advances in Information Technology,* vol. 14, no. 5, pp. 1132–1142. doi: 10.12720/jait.14.5.1132-1142
55. S. Jha, et al. "Neutrosophic approach for enhancing quality of signals," *Multimedia Tools and Applications*, 2019, doi: 10.1007/s11042-019-7375-0

8 Quantization of Artificial Intelligence Models for Personal Protective Equipment Detection in Industrial Environments

Khoa Nguyen Dang, Vijender Kumar Solanki, Tran Thi Van, Pham Tuan Minh, and Duc-Tan Tran

8.1 INTRODUCTION

Personal Protective Equipment (PPE) plays a critical role in industrial environments by protecting workers from a variety of hazards and ensuring their safety and well-being [1, 2]. These hazards may include physical dangers, chemical exposures, electrical risks, and biological threats. By serving as a protective barrier, PPE helps mitigate potential risks present in the workplace.

Typically, a manager stationed at the entrance of each factory or construction site is responsible for inspecting workers' protective gear. However, this manual process carries risks of inefficiency, potential fraud, and safety concerns. With advancements in science and technology, this task has been increasingly automated. Two primary approaches have been proposed to address the issue: traditional vision processing [3] and more advanced methods based on artificial intelligence (AI) [4].

Traditional vision processing involves comparing and matching the current image with a stored image using algorithms such as Scale-Invariant Feature Transform (SIFT) [5], Speeded-Up Robust Features (SURF) [6, 7], Robust Independent Elementary Features (BRIEF) [8], and Rotated BRIEF (ORB) [9]. These algorithms excel in recognizing protective gear with high speed [10]. However, their performance is heavily reliant on factors like camera quality, camera angle, and environmental conditions. As a result, there is a need for alternative methods, such as AI, to address these limitations and enhance reliability.

In recent years, numerous AI-based solutions have emerged for PPE detection, including Convolutional Neural Networks (CNNs) [11, 12] with an accuracy of 79%, Fast Region-based CNN [13], and MobileNet [14]. While these methods offer effective detection of protective gear for workers, they still face challenges related to

DOI: 10.1201/9781003609773-8

real-time performance, accuracy, and handling objects at different scales. To address these shortcomings, newer versions of the YOLO series, such as YOLOv3 [15] with 90% accuracy, YOLOv4 [16], YOLOv5 [17], and YOLOv7 [18], have gained significant recognition in this domain. These models are continuously updated and made available to users and developers, enabling their application in research, practical solutions, and specialized industries.

AI algorithms have achieved remarkable success when deployed on computers, high-performance computing (HPC) systems, and cloud platforms. However, building generalized systems is not always feasible due to constraints such as environmental conditions, workspace limitations, and costs. In practical applications, embedded computers, microcontrollers, and 8-bit microchips are often employed during the development process. AI models designed for 32-bit or 64-bit systems cannot run directly on these resource-constrained devices. Additionally, challenges related to limited memory and storage capacity hinder the deployment of full-featured models. To address these issues, quantization [19–21] is utilized, which maps a large set of values to a smaller set. This technique is commonly applied in digital signal processing and machine learning to optimize model performance for constrained environments.

In this research, we utilize YOLOv8 to detect PPE worn by workers in factories and industrial settings. The model is quantized by converting the floating-point variables in the AI model to smaller integer types. The quantized model is then validated on the Jetson Nano, a widely adopted platform for deploying AI models in real-world applications. The results demonstrate that both the original YOLOv8 and the quantized version achieve comparable performance in tracking PPE, maintaining an accuracy rate exceeding 80% across multiple objects and varying viewing angles.

The structure of the chapter is organized as follows: Section 8.2 presents the overall system configuration, while Section 8.3 provides an introduction to YOLOv8. Section 8.4 details the application and explanation of the quantization algorithm. The results are analyzed and discussed in Section 8.5. Finally, Section 8.6 summarizes the key findings and conclusions.

8.2 OVERALL CONFIGURATION OF THE SYSTEM

The processing is carried out across four subsystems: the training model, the conversion to the Open Neural Network Exchange (ONNX) model, the quantization model, and inference on a device, as illustrated in Figure 8.1. Initially, the training model is executed using Google Colab [22], with the dataset being utilized to generate the AI model using the PyTorch library.

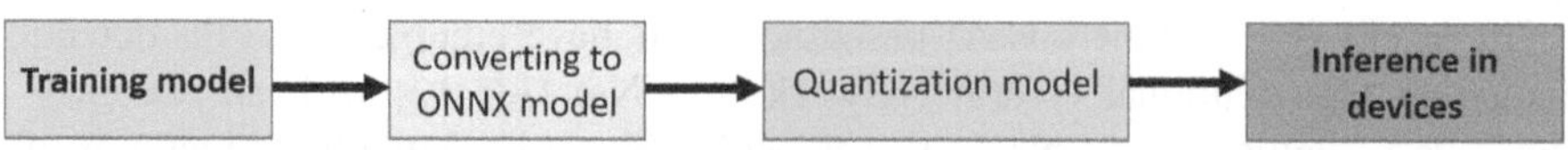

FIGURE 8.1 Overall configuration of the system.

Next, the AI model generated in Google Colab is converted into the ONNX format [23]. ONNX serves as an intermediary model platform that supports conversions across various frameworks, including Keras, TensorFlow, Scikit-Learn, PyTorch, and XGBoost.

Third, the quantization process converts the AI model from 32-bit floating-point precision (FP32) to 8-bit integer (INT8), enabling it to run on simple microcontrollers such as ARM, PIC, AVR, and others. This conversion serves multiple purposes: reducing the model size, minimizing resource requirements like memory and energy consumption, and speeding up inference, as computations with INT8 are typically faster than FP32 on optimized hardware.

Fourth, inference on a device enables the INT8 AI model to run using specialized libraries such as TensorRT and ONNX Runtime, making it applicable for real-world applications. In this step, inputs such as camera feeds, images, and data from other sensors are utilized as inputs for the model.

In this research, we develop a system to predict the usage of PPE in industrial plants, aiming to reduce risks and enhance worker safety. To achieve this, YOLOv8 is employed and further analyzed in the subsequent sections.

8.3 YOLOV8 STRUCTURE

YOLOv8 is structured into three main components: Backbone, Neck, and Head [24], as illustrated in Figure 8.2. First, the Backbone is responsible for extracting image features. This step typically employs convolutional layers, batch normalization, and activation functions to reduce the spatial dimensions of the image while increasing the number of feature channels. This process is based on the cross-stage partial (CSP) method.

Next, the Neck component processes multi-scale features and enhances them using techniques such as upsampling, downsampling, channel concatenation, convolution, and a CSP Bottleneck with a 2-fused path (Cf2). This step ensures that features at different scales are effectively captured and refined.

Finally, the Head component, which includes the prediction head, classification, and loss computation, utilizes convolutional modules (ConvModule and Conv2d) to extract feature outputs and perform regression (Bbox_Loss) and classification loss (Cls_Loss). This step integrates all the information from the Neck, producing predictions for object positions across multiple scales, classifications, and loss calculations.

8.4 DEVELOPMENT OF A QUANTIZATION ALGORITHM

Quantization is a technique that reduces the time required to create an AI model while maintaining acceptable accuracy. It works by converting floating-point variables used in operators and representing the algebraic objects of the vector space, defined as tensors, into smaller data types such as integers (INT) or 8-bit integers (INT8). This process helps reduce the model's size and computational demands. The quantization workflow is illustrated in Figure 8.3.

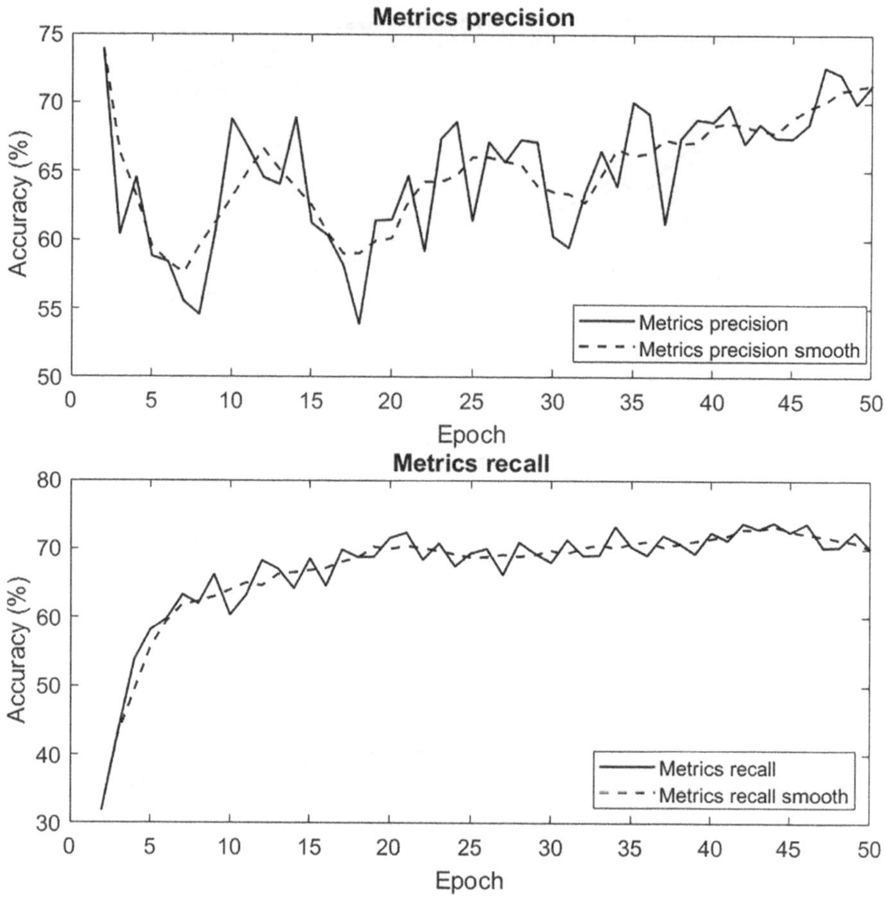

FIGURE 8.6 Metric in the training step. (a) Metric precision. (b) Metric recall.

the original YOLOv8n, as evidenced by metrics such as mAP, Precision, Recall, and F1-score.

Ultimately, an AI model is developed and can be deployed in real-world applications. By utilizing data from a camera, the AI model predicts whether workers are properly equipped with protective gear, such as vests and gloves. The results are displayed on a monitoring system, and alerts are generated for individuals who do not meet the required safety standards.

The YOLOv8n model, after being quantized, is deployed on a device called the Jetson Nano [26], a compact AI-enabled computer that supports high-performance execution of AI models. The hardware configuration is illustrated in Figure 8.9. A camera serves as the input device, connected to the Jetson Nano via a USB interface, with its specifications detailed in Table 8.2. Following the prediction process, the worker's status is displayed on a monitor and, if the protective gear does not

```
Class               images   instances   Box(P)      R    mAP50    mAP50-95
all                    315        2261    0.730  0.722    0.739       0.497
NO glove                57         102    0.655  0.373    0.467       0.182
NO hardhat              83         128    0.796  0.672    0.713       0.329
NO safety shoe          41          85    0.527  0.812    0.574       0.413
NO safety vest          73          78    0.688  0.393    0.523       0.317
glove                  207         400    0.660  0.568    0.581       0.246
hardhat                184         240    0.771  0.683    0.821       0.341
person                 276         378    0.871  0.761    0.905       0.812
safety shoe            198         450    0.736  0.880    0.876       0.648
safety vest            291         450    0.863  0.956    0.963       0.791

Speed: 2.5ms preprocess, 5.2ms inference, 0.0ms loss, 7.2ms postprocess per image

=== Result ===
mAP50 = 0.7399
mAP50-95 = 0.4975
Precision = 0.7023
Recall = 0.7201
F1-score = 0.7401
```

FIGURE 8.7 Testing results of the model without quantization.

```
Class               Images   Instances   Box(P)      R    mAP50    mAP50-95
all                    315        2261    0.730  0.722    0.719       0.487
NO glove                57         102    0.655  0.373    0.467       0.182
NO hardhat              83         128    0.796  0.672    0.713       0.329
NO safety shoe          41          85    0.527  0.812    0.574       0.413
NO safety vest          73          78    0.688  0.393    0.523       0.317
glove                  207         400    0.660  0.568    0.581       0.246
hardhat                184         240    0.771  0.683    0.821       0.341
person                 276         378    0.871  0.761    0.905       0.812
safety shoe            198         450    0.736  0.880    0.876       0.648
safety vest            291         450    0.863  0.956    0.963       0.791

Speed: 2.1ms preprocess, 5.0ms inference, 0.0ms loss, 2.4ms postprocess per image
=== Result ===
mAP50       : 0.7191
mAP50-95    : 0.4871
Precision   : 0.7296
Recall      : 0.7220
F1-score    : 0.7258
```

FIGURE 8.8 Testing results of the model with quantization.

meet the required standards, the information is sent to warning devices for further action.

In this experiment, we examine two primary scenarios: one involving a single individual and the other involving multiple people. In each scenario, the workers are either equipped with protective gear or not.

As shown in Figure 8.10, a single person is detected in the camera view with an accuracy of 88%, and the system determines that the individual is not wearing any protective gear (25% for a vest and 53% for a hat). Additionally, in Figure 8.11, the

TABLE 8.1
Comparison between YOLOv8 Model and YOLOv8n-Quantization Model

Parameters	Original YOLOv8n	YOLOv8n-Quantization
mAP50	73.39	71.91
mAP50-95	49.75	48.71
Precision	73.7	72.86
Recall	73.92	72.2
F1-score	74.01	72.58
Speed preprocess	2.5 ms	2.1 ms
Inference	5.2 ms	5.0 ms
Postprocess per image	7.2 ms	2.4 ms

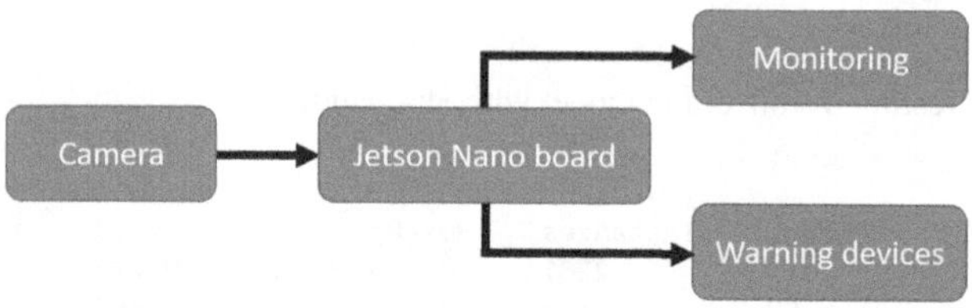

FIGURE 8.9 Testbed AI model in embedded devices.

TABLE 8.2
Specification of the Camera

Name	Specifications
Camera name	A850 integrated microphone
Connection	USB 2.0
Speed	30 fps
Supports	Video HD 720p

system predicts the presence of the person with accuracies of 71% and 79%, the vest with 91% and 95%, and the hat with 43% and 66%. These results demonstrate the system's performance when the model is executed on an embedded device using quantization.

In another scenario, many people in the camera view are performed as in Figure 8.11. In case more than one person is in camera view, the AI model still provides high performance in detecting protective gear, as in Figure 8.12(a) and 8.12(b). In (a), one person with no protection is predicted and the other with safety. In another case, Figure 8.12(b), all workers are supported with all protection. Lastly, all results are proof of the performance of the method in the real application of industry

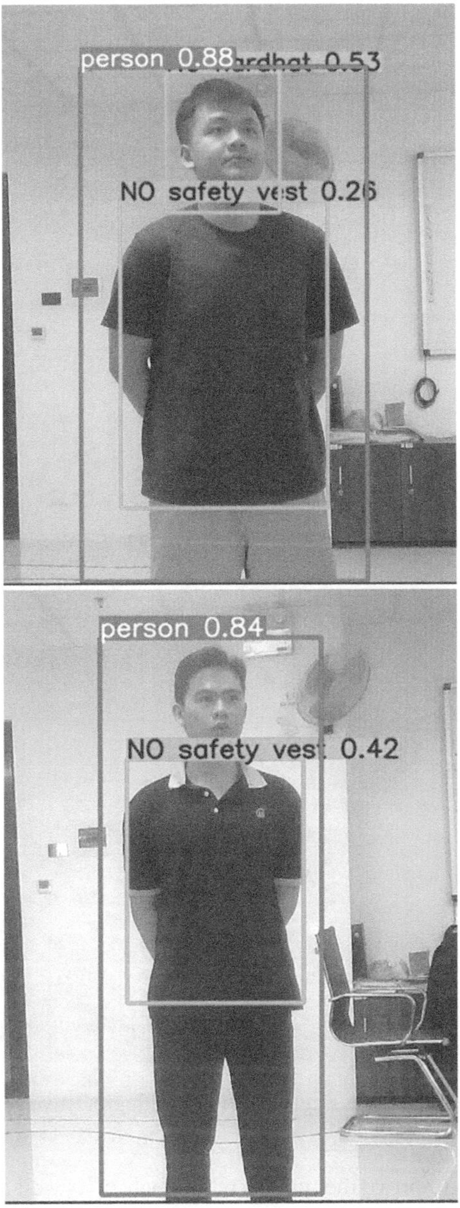

FIGURE 8.10 Camera view of a single individual without protective gear.

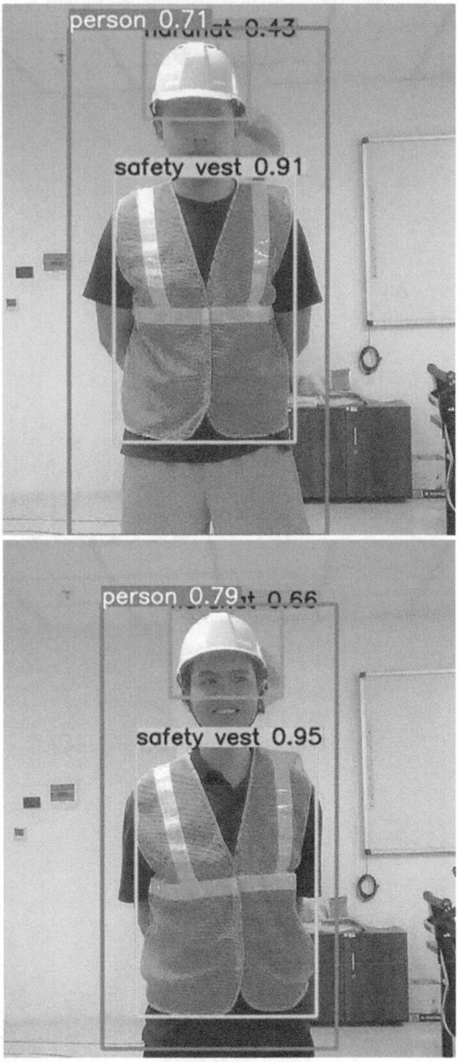

FIGURE 8.11 Camera view of a single individual with protective gear.

that could be used to improve the quality in the factory as well as reduce risk in labor and production.

8.6 CONCLUSION

This chapter introduces a quantization method to optimize the YOLOv8 model for detecting PPE in industrial settings by converting float-type variables to INT-type. The generated model satisfies the requirement of accurately identifying whether the protective gear is worn, achieving an accuracy of over 80%. Additionally, the model is

FIGURE 8.12 Camera view of many people. (a) One of the people with no protective gear. (b) Full protective gear.

validated on Jetson Nano, an embedded computing board. All results demonstrate the effectiveness of the proposed method, highlighting its potential as a robust solution not only for this specific application but also for addressing quantization challenges in general.

REFERENCES

[1] S. Rasouli, Y. Alipouri, and S. Chamanzad, "Smart Personal Protective Equipment (PPE) for Construction Safety: A Literature Review," *Saf. Sci.*, vol. 170, 106368, Feb. 2024, doi: 10.1016/j.ssci.2023.106368.

[2] A. Elesawy, E. Mohammed Abdelkader, and H. Osman, "A Detailed Comparative Analysis of You Only Look Once-Based Architectures for the Detection of Personal Protective Equipment on Construction Sites," *Eng*, vol. 5, no. 1, pp. 347–366, 2024, doi: 10.3390/eng5010019.

[3] A. M. Vukicevic, M. Petrovic, P. Milosevic, A. Peulic, K. Jovanovic, and A. Novakovic, "A Systematic Review of Computer Vision-Based Personal Protective Equipment Compliance in Industry Practice: Advancements, Challenges and Future Directions," *Artif. Intell. Rev.*, vol. 57, no. 12, p. 319, Oct. 2024, doi: 10.1007/s10462-024-10978-x.

[4] Q. Hu, X. Shen, X. Qian, G. Huang, and M. Yuan, "The Personal Protective Equipment (PPE) Based on Individual Combat: A Systematic Review and Trend Analysis," *Def. Technol.*, vol. 28, pp. 195–221, Oct. 2023, doi: 10.1016/j.dt.2022.12.007.

[5] D. Pennati and L. Bocchi, "Analysis of the Relationship Between Scale Invariant Feature Transform Keypoint Properties and Their Invariance to Geometrical Transformation Applied to Cone-Beam Computed Tomography Images," *Bioengineering*, vol. 11, no. 12, 2024, doi: 10.3390/bioengineering11121236.

[6] N. Zhang, "Computing Parallel Speeded-Up Robust Features (P-SURF) via POSIX Threads," in *Emerging Intelligent Computing Technology and Applications*, D.-S. Huang, K.-H. Jo, H.-H. Lee, H.-J. Kang, and V. Bevilacqua, Eds., Berlin, Heidelberg: Springer Berlin Heidelberg, 2009, pp. 287–296.

[7] A. Setiawan, R. A. Yunmar, and H. Tantriawan, "Comparison of Speeded-Up Robust Feature (SURF) and Oriented FAST and Rotated BRIEF (ORB) Methods in Identifying Museum Objects Using Low Light Intensity Images," *IOP Conf. Ser. Earth Environ. Sci.*, vol. 537, no. 1, 012025, Jul. 2020, doi: 10.1088/1755-1315/537/1/012025.

[8] P. Kaur and N. Kumar, "BRIEFBCS: Binary Robust Independent Elementary Features Based Fuzzy Vault Scheme in BCS," *Clust. Comput.*, vol. 27, no. 3, pp. 3135–3148, Sep. 2023, doi: 10.1007/s10586-023-04134-3.

[9] N. R. Nusantika, J. Xiao, and X. Hu, "An Enhanced Multiscale Retinex, Oriented FAST and Rotated BRIEF (ORB), and Scale-Invariant Feature Transform (SIFT) Pipeline for Robust Key Point Matching in 3D Monitoring of Power Transmission Line Icing with Binocular Vision," *Electronics*, vol. 13, no. 21, 2024, doi: 10.3390/electronics13214252.

[10] M. Riaz, S. Ahmed, F. Khan, and J. Lee, "Enhancing Workplace Safety: PPE_Swin—A Robust Swin Transformer Approach for Automated Personal Protective Equipment Detection," *Electronics*, vol. 12, no. 22, p. 4675, 2023, doi: 10.3390/electronics12224675.

[11] F. Zhafran, E. S. Ningrum, M. N. Tamara, and E. Kusumawati, "Computer Vision System Based for Personal Protective Equipment Detection, by Using Convolutional Neural Network," in 2019 International Electronics Symposium (IES), Sep. 2019, pp. 516–521. doi: 10.1109/ELECSYM.2019.8901664.

[12] G. Gallo, F. D. Rienzo, F. Garzelli, P. Ducange, and C. Vallati, "A Smart System for Personal Protective Equipment Detection in Industrial Environments Based on Deep Learning at the Edge," *IEEE Access*, vol. 10, pp. 110862–110878, 2022, doi: 10.1109/ACCESS.2022.3215148.

[13] M. I. Ahmed, M. Rahman, S. Khan, and H. Kim, "Personal Protective Equipment Detection: A Deep-Learning-Based Sustainable Approach," *Sustainability*, vol. 15, no. 18, 13990, 2023, doi: 10.3390/su151813990.

[14] S. Al-Azani, H. Luqman, M. Alfarraj, A. A. I. Sidig, A. H. Khan, and D. Al-Hamed, "Real-Time Monitoring of Personal Protective Equipment Compliance in Surveillance Cameras," *IEEE Access*, vol. 12, pp. 121882–121895, 2024, doi: 10.1109/ACCESS.2024.3451117.

[15] S. Qiu and Y. Kang, "A Personal Protective Equipment Safety Monitoring System in Industrial Scenes," in 2022 IEEE 8th International Conference on

Computer and Communications (ICCC), Dec. 2022, pp. 2182–2186. doi: 10.1109/ICCC56324.2022.10065769.

[16] A. A. Protik, A. H. Rafi, and S. Siddique, "Real-time Personal Protective Equipment (PPE) Detection Using YOLOv4 and TensorFlow," in 2021 IEEE Region 10 Symposium (TENSYMP), Aug. 2021, pp. 1–6. doi: 10.1109/TENSYMP52854.2021.9550808.

[17] L. Yipeng and W. Junwu, "Personal Protective Equipment Detection for Construction Workers: A Novel Dataset and Enhanced YOLOv5 Approach," *IEEE Access*, vol. 12, pp. 47338–47358, 2024, doi: 10.1109/ACCESS.2024.3382817.

[18] Z. Wang, Y. Zhu, Y. Zhang, and S. Liu, "An Effective Deep Learning Approach Enabling Miners' Protective Equipment Detection and Tracking Using Improved YOLOv7 Architecture," *Comput. Electr. Eng.*, vol. 123, 110173, Apr. 2025, doi: 10.1016/j.compeleceng.2025.110173.

[19] B. Jacob, S. Kligys, B. Chen, M. Zhu, M. Tang, A. Howard, H. Adam, and D. Kalenichenko, "Quantization and Training of Neural Networks for Efficient Integer-Arithmetic-Only Inference," in 2018 IEEECVF Conf. Comput. Vis. Pattern Recognit., pp. 2704–2713, 2017.

[20] L. Wei, Z. Ma, C. Yang, and Q. Yao, "Advances in the Neural Network Quantization: A Comprehensive Review," *Appl. Sci.*, vol. 14, no. 17, 2024, doi: 10.3390/app14177445.

[21] B. Rokh, A. Azarpeyvand, and A. Khanteymoori, "A Comprehensive Survey on Model Quantization for Deep Neural Networks in Image Classification," *ACM Trans Intell Syst Technol*, vol. 14, no. 6, Nov. 2023, doi: 10.1145/3623402.

[22] K. Edwards, C. Scalisi, J. DeMars-Smith, and K. Lee, "Google Colab for Teaching CS and ML," in Proceedings of the 55th ACM Technical Symposium on Computer Science Education V. 2, *in SIGCSE 2024*. New York, NY, USA: Association for Computing Machinery, 2024, p. 1925. doi: 10.1145/3626253.3635432.

[23] F. Manca, F. Ratto, and F. Palumbo, "ONNX-To-Hardware Design Flow for Adaptive Neural-Network Inference on FPGAs," in Embedded Computer Systems: Architectures, Modeling, and Simulation, L. Carro, F. Regazzoni, and C. Pilato, Eds., Cham: Springer Nature Switzerland, 2025, pp. 85–96.

[24] B.-U. Shin, "Development of YOLOv8 and Segment Anything Model Algorithm-Based Hanok Object Detection Model for Sustainable Maintenance of Hanok Architecture," *Sustainability*, vol. 16, no. 9, 2024, doi: 10.3390/su16093775.

[25] S. Policepatil, J. Sharma, B. Kumar, D. Singh, S. Pramanik, A. Gupta, and M. B. S. Basha, "Financial sector hyper-automation: Transforming banking and investing procedures," in M. S. J. Justin, R. Jalagat, K. Chandar, P. Aquino, and K. Sayari, Eds., *Examining global regulations during the rise of fintech*, Pennsylvania: United States, IGI Global, 2025, pp. 299–318.

[26] C. Choe, M. Choe, and S. Jung, "Run Your 3D Object Detector on NVIDIA Jetson Platforms: A Benchmark Analysis," *Sensors*, vol. 23, no. 8, 2023, doi: 10.3390/s23084005.

9 Frequency and Spatial Dual-Domain Filtering Scheme for Viscoelasticity Imaging in Ultrasound Elastography

Tran Quang Huy, Luong Thi Theu, Vijender Kumar Solanki, and Duc-Tan Tran

9.1 INTRODUCTION

Shear wave viscoelasticity imaging (SWEI) is a pivotal technique in the field of medical diagnostics, enabling the noninvasive assessment of the mechanical properties of biological tissues. These properties, which include elasticity and viscosity, serve as critical indicators of tissue health and pathology [1]. SWEI operates by analyzing the propagation of shear waves through tissues, thereby providing quantitative insights into the complex shear modulus (CSM) of the medium [2]. This allows for the detection and characterization of tissue abnormalities [3]. However, the accurate reconstruction of CSM images poses significant challenges due to the intrinsic noise in shear wave measurements. This issue is particularly pronounced in heterogeneous and complex tissue environments, where variations in tissue composition and structure can obscure essential diagnostic details [4]. To address these challenges, advanced imaging techniques and noise reduction strategies must be employed, ensuring that the resulting CSM images are accurate and reliable for clinical use [5]. SWEI has evolved to become an essential tool in the arsenal of medical imaging, offering unparalleled insights into the viscoelastic properties of tissues. By enhancing the quality of CSM imaging, we can improve the accuracy of diagnoses and ultimately contribute to better patient outcomes [6]. As research in this field continues to advance, the development of innovative methodologies to minimize noise and enhance image clarity remains a top priority [7]. This ongoing effort will ensure that SWEI maintains its crucial role in the noninvasive evaluation of tissue mechanics, supporting the early detection and treatment of various medical conditions [8].

Sarvazyan et al. initially introduced the concept of shear wave elasticity imaging (SWEI) for medical diagnostics in 1998 [9]. Building on this foundation, Chen and

DOI: 10.1201/9781003609773-9

colleagues formulated a quantitative method in 2004 to determine tissue elasticity and viscosity through shear wave velocity measurements, deriving an equation that links shear wave speed to the medium's mechanical properties [10]. Subsequently, in 2010, Orescanin Marko et al. employed the Kelvin-Voigt model to develop the Maximum Likelihood Ensemble Filter (MLEF) for estimating CSM parameters in homogeneous media [11]. This approach was further refined in 2013 to adapt MLEF for heterogeneous environments, allowing for more accurate CSM estimation in complex tissues [12]. Despite its effectiveness, the MLEF methodology demands significant computational power, posing challenges for real-time imaging. Moreover, a multitude of research teams have continued to explore CSM estimation and advancements in shear wave imaging [13–17].

Traditional elastic ultrasonic imaging systems typically function at a single fixed frequency, employing low-pass filtering (LPF) techniques to mitigate high-frequency noise [18]. Although these methods are effective in isolating the desired signal, they are often inadequate for heterogeneous media where noise spans multiple frequency bands, potentially obscuring critical diagnostic information. Consequently, the resultant images may exhibit spatial noise and diminished contrast, compromising their diagnostic precision. To address this limitation, recent research has focused on developing adaptive filtering methods [19] and sophisticated algorithms designed to enhance the accuracy of imaging in heterogeneous tissues. These approaches include multi-frequency filtering techniques that can better differentiate between noise and the true signal, thus improving image clarity and contrast. However, the increased computational complexity associated with these advanced methods poses significant challenges for their implementation in real-time imaging applications.

The propagation of shear waves in biological soft tissues is frequently modeled through particle velocity and its historical values are linked by the fundamental wave equation. This model, however, is only appropriate for homogeneous media. In heterogeneous media such as biological soft tissues, shear wave velocity is influenced not only by its historical values but also by values at adjacent locations. Therefore, in this chapter, we employ the finite-difference time-domain (FDTD) method to simulate the propagation of shear waves in biological tissues, which are inherently complex and heterogeneous. The findings demonstrate that shear wave particle velocity is dependent on both spatial and temporal variables. Furthermore, the particle velocity data derived from shear wave measurements is often contaminated with noise from the measurement process. To mitigate this issue, we propose an innovative dual-filter technique. This method combines band-pass filtering (BPF) in the frequency domain with median filtering in the spatial domain, aiming to enhance the quality of CSM images. The integration of these two filtering strategies allows us to capitalize on the strengths of both frequency and spatial domain filtering. In essence, the BPF isolates the excitation frequency and effectively minimizes both high- and low-frequency noise, while the median filter is applied to the spatial data to further reduce noise, preserving critical structural details. Our results indicate that this two-filter approach substantially reduces normalized errors and significantly improves image clarity. This dual-filter strategy presents an efficient solution for real-time CSM imaging, offering a robust method for the enhanced imaging of viscoelastic properties. Consequently, it holds considerable potential for various clinical applications where accurate real-time tissue characterization is crucial.

Our methodology commences with a FDTD simulation to model the propagation of shear waves in heterogeneous tissues. This is followed by the application of a BPF that effectively isolates the excitation frequency while minimizing the high- and low-frequency noise present in the shear wave data. The Algebraic Helmholtz Inversion (AHI) algorithm is then employed to estimate the CSM from the filtered data. Subsequently, a median filter is utilized on the spatial data to further diminish noise levels, ensuring the preservation of critical structural details within the tissue images. This dual-filter strategy leverages the benefits of both frequency and spatial domain filtering techniques. The BPF component is crucial for isolating the desired signal by reducing noise that spans across multiple frequency bands, thus enhancing the clarity and contrast of the images. Concurrently, the median filter, applied in the spatial domain, is instrumental in preserving the integrity of structural boundaries and the contrast between different tissue regions. The efficacy of our proposed method was validated through extensive numerical simulations, which assessed its performance in terms of noise reduction and the accuracy of elasticity and viscosity estimates. The results from these simulations indicated that the two-filter approach markedly reduced normalized errors and significantly improved image clarity. This comprehensive strategy provides an efficient solution for real-time CSM imaging, offering a robust method for the enhanced visualization of viscoelastic properties in biological tissues. In short, our work presents a significant advancement in shear wave elastography, proposing a novel dual-filter technique that enhances the accuracy and reliability of real-time imaging. This methodology holds substantial potential for various clinical applications, where precise and timely characterization of tissue properties is essential for effective diagnosis and treatment planning.

9.2 METHODOLOGY

In the system designed forSWEI, oscillations along the Z-axis are induced by a needle vibrating at a set frequency, producing shear waves that propagate within the X-Y plane. The generated shear waves are then measured using Doppler ultrasound devices to capture the resulting particle velocities, providing crucial data for analysis [20]. The needle's controlled oscillation generates consistent shear waves, which traverse the soft tissue in predictable patterns. These waves are crucial for understanding the mechanical properties of the tissue, as they allow for the detailed mapping of viscoelastic characteristics. By employing Doppler ultrasound, the system accurately measures the minute velocities of particles within the tissue, which are influenced by the shear waves' propagation. This detailed particle velocity data is essential for creating high-fidelity images that reveal the intricate details of tissue structure and composition. The ability to measure and analyze these velocities in real-time offers significant advantages for clinical diagnostics, enabling precise and timely assessments of tissue health and pathology.

The experimental setup for shear wave excitation and measurement is shown in Figure 9.1.

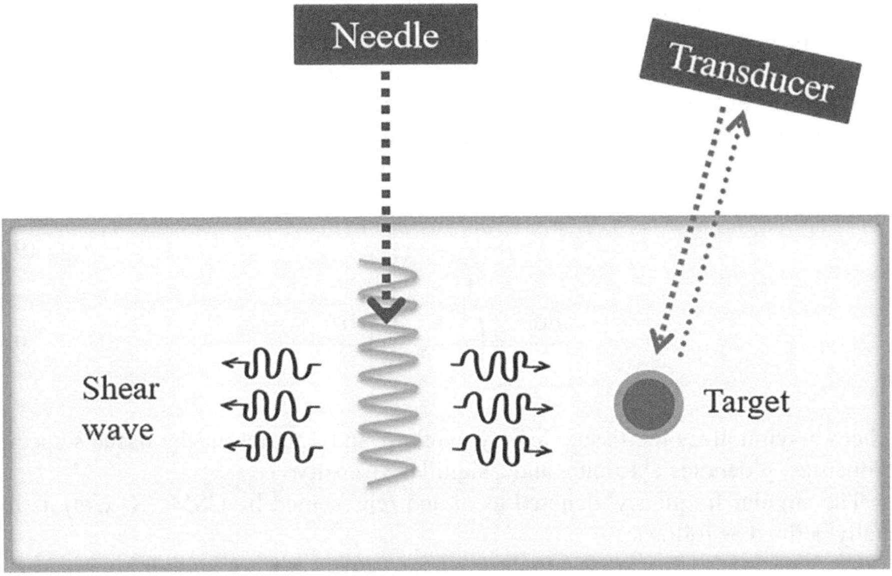

FIGURE 9.1 Excitation and measurement of shear waves using a vibrating needle. The experimental setup illustrates the generation of shear waves by mechanical excitation and the corresponding measurement configuration used for wave propagation and displacement acquisition.

Equation (9.1) is utilized to calculate the particle velocity $\vartheta(a,\tau)$ at a particular spatial position a and time τ [20]:

$$\vartheta(a,\tau)=\frac{A}{e^{\gamma(a-a_0)}\sqrt{a-a}}cos\left[\omega\tau-\kappa_s\left(a-a_0\right)-\varphi\right] \tag{9.1}$$

In this context, A denotes the amplitude of the needle's vibration, while $a_0$indicates its spatial location. The attenuation coefficient and wavenumber at a specific spatial position are represented by γ and κ_s, respectively. Additionally, φ signifies the initial phase.

Equation (9.1) simplifies the calculation of particle velocity, which is advantageous, but it fails to take into account the propagation of shear waves in real tissue, particularly in heterogeneous environments. To address this shortcoming, the FDTD method is employed. This method assumes that shear wave propagation follows the radial axis, and it disregards medium absorption. Equations (9.2) and (9.3) detail the relationship between the particle velocity vectorϑ_z, which is aligned with the x-axis in the direction of wave propagation within the Cartesian coordinate system, and the compression tensor $\delta\delta_{zx}$ [21]. Specifically, Equation (9.2) represents the time derivative of the particle velocity as a function of the spatial derivative of the compression tensor, while Equation (9.3) describes the spatial derivative of the compression tensor as being dependent on the tissue's elastic and viscous properties, denoted by ϱ and ς, respectively. By employing these equations, researchers can better discretize the complex interactions of shear waves within tissue, leading to more accurate representations and analyses in shear wave elastography. This refined approach enhances the fidelity of diagnostics, providing a clearer understanding of the mechanical properties of soft tissues and ultimately improving patient outcomes.

$$\rho \frac{\partial v_z}{\partial_\tau} = \frac{\partial \delta_{zx}}{\partial \tau}, \tag{9.2}$$

$$\frac{\partial \delta_{zx}}{\partial \tau} = \left(\varrho + \varsigma \frac{\partial}{\partial \tau} \right) \frac{\partial \vartheta_z}{\partial_\tau} \tag{9.3}$$

where ρ symbolizes the tissue density, while ϱ and ζ represent the tissue's specific properties. ϱ denotes elasticity, and ζ signifies viscosity.

The angular frequency, denoted as ω and represented by CSM, $\mathrm{X}(x, \omega)$ is formally defined as follows:

$$\mathrm{X}(x, \omega) = \varrho(x) - i\omega\varsigma(x) \tag{9.4}$$

where ϱ stands for the elasticity of the tissue, while ς is used to symbolize the viscosity that needs to be determined. To be more specific, ϱ represents how the tissue can return to its original shape after being deformed, which is a measure of its elasticity. On the other hand, ς indicates the tissue's resistance to flow or change in shape, which is a measure of its viscosity. Both of these properties are crucial for understanding the behavior of the tissue under different conditions.

The following symbols are utilized to discretize Equations (9.2) and (9.3):

$$\vartheta_z(x, \tau) = \vartheta_z(i\Delta x, n\Delta\tau) = \vartheta_z^n\Big|_i \tag{9.5}$$

$$\delta_{zx}(x, \tau) = \delta_{zx}(i\Delta x, n\Delta\tau) = \delta_{zx}^n\Big|_i \tag{9.6}$$

To proceed, we define the indices appropriately: let i represent the spatial index and n denote the temporal index. The parameter Δx is designated as the distance between successive spatial locations, whereas $\Delta\tau$ stands for the sampling time interval. These definitions will facilitate the understanding and implementation of the FDTD approach that is used to solve the relevant equations. The spatial index i helps in mapping out the different spatial points of interest, while the temporal index n is crucial for tracking changes over time. By setting Δx as the fixed spatial increment, we ensure consistent spacing between these points, which is essential for accurate modeling. Similarly, defining $\Delta\tau$ as the time interval helps in maintaining uniform time steps throughout the simulation, allowing for precise temporal resolution. Together, these parameters create a structured framework that aids in the effective discretization and solution of the equations at hand.

The FDTD approach is employed to numerically solve Equations (9.2) and (9.3) by discretizing the equations and applying the appropriate indices for space and time. This method allows for the precise calculation of wave propagation through a medium by breaking down the equations into discrete time and space steps. In more detail, the FDTD method involves defining the spatial index i and the temporal index n, with Δx representing the distance between successive spatial locations and Δτ denoting the time interval between successive temporal points. These definitions are crucial for accurately mapping the spatial and temporal domains over which the equations are solved. By discretizing the equations, the FDTD method transforms the continuous equations into a series of algebraic equations that can be solved iteratively. This approach is particularly useful for simulating complex wave interactions in tissues, as it allows for detailed analysis of how waves propagate through the medium. By applying the FDTD method, we can gain insights into the mechanical properties of the tissue, such as elasticity and viscosity, and how these properties influence wave propagation.

$$\vartheta_z^{n+1}\Big|_i = \vartheta_z^{n}\Big|_i + \frac{\Delta\tau}{\rho\Delta x}\left(\delta_{zx}^{n+\frac{1}{2}}\Big|_{i+\frac{1}{2}} - \delta_{zx}^{n+\frac{1}{2}}\Big|_{i-\frac{1}{2}}\right) \tag{9.7}$$

$$\delta_{zx}^{n+\frac{1}{2}}\Big|_{i+\frac{1}{2}} = \delta_{zx}^{n-\frac{1}{2}}\Big|_{i+\frac{1}{2}} + \frac{\varrho\Delta\tau}{\Delta x}\left(\vartheta_z^{n+1}\Big|_{i+1} - \vartheta_z^{n+1}\Big|_i\right) + \frac{\varsigma}{\Delta x}\left(\vartheta_z^{n+1}\Big|_{i+1} - \upsilon_z^{n+1}\Big|_i\right) - \frac{\varsigma}{\Delta x}\left(\vartheta_z^{n}\Big|_{i+1} - \vartheta_z^{n}\Big|_i\right) \tag{9.8}$$

Measurement errors and reflections can affect particle velocity. To improve signal quality before using AHI for CSM estimation, we applied a BPF to reduce noise in the data from Doppler ultrasound equipment. Unlike the low-pass filter used in [22], which only removes high-frequency noise, our method retains the excitation frequency range.

The particle velocity signal, $\vartheta_z(n)$, is sent into the tissue and affected by noise, $\theta(n)$. The resulting noisy signal $\vartheta(n)$ is given as:

$$\vartheta(n) = \vartheta_z(n) + \theta(n) \tag{9.9}$$

The noisy signal, $\vartheta(n)$, is input into a BPF to produce the required signal estimate, $\vartheta_z(n)$. The filtered signal, $\hat{\vartheta}_z(n)$, is then sampled into an N-sample vector:

$$\vartheta(n) = \left[\vartheta_z(0)\ \vartheta_z(1)\ \ldots\ \vartheta_z(N-1)\right], \tag{9.10}$$

and the filter coefficients are calculated as:

$$\omega(n) = \left[\omega(0)\ \omega(1)\ \ldots\ \omega(\Gamma)\right] \tag{9.11}$$

here Γ is the order of the filter, while the weights $\omega(n)$ represent the filter coefficients.

Assuming isotropic viscous elasticity for small tissue volumes, combining Equations (9.2) and (9.3) yields:

$$\rho\frac{\partial^2 \vartheta_z}{\partial t^2} = \mathrm{X}(x,t)\nabla^2 v_z \tag{9.12}$$

in this case, $\mathrm{X}'(x,t)$ represents the CSM in the time domain, while $\nabla^2 \vartheta_z$ represents the Laplacian of ϑ_z, which is computed as $\nabla^2 \vartheta_z = \frac{\partial^2 \vartheta_z}{\partial x^2}$.

The algorithm for the AHI mentioned in reference [22] is employed to solve the given Equation (9.12). By applying this sophisticated algorithm, Equation (9.12) is transformed and simplified into what is known as the Helmholtz Equation, providing a more manageable form for analysis and computation. This step is crucial in computational simulations and helps in accurately determining the behavior of the system under study. The transformation into the Helmholtz Equation allows for the use of various numerical methods to find solutions that are otherwise difficult to obtain.

$$\left(\frac{\mathrm{X}(x,\omega)}{\rho}\nabla^2 + \omega^2\right)\Omega_z(x,\omega)\Big|_{\omega=\omega_0} = 0 \tag{9.13}$$

here $\mathrm{X}(x,\omega)$ represents the CSM in the frequency domain, as defined in Equation (9.4). The Fourier transform of the particle velocity $\vartheta_z(x,t)$ with respect to time is denoted by $\Omega_z(x,\omega)$, where $\Omega_z(x,\omega) = F_t\{\vartheta_z(x,t)\}$. Additionally ω_0 represents the angular frequency, which is given by $\omega_0 = 2\pi f_0$.

To delve deeper, the CSM can be derived by performing the Fourier transform on the particle velocity data acquired over a period of time, converting it into the frequency domain. This transformation is essential as it allows for a more detailed analysis of the shear moduli's response to varying frequencies. In Equation (9.13), the relationship between the CSM and the particle velocity is clearly illustrated. The angular frequency, denoted as ω_0, is integral in this process, serving as a fundamental parameter that influences the behavior and characteristics of the resulting CSM. By applying the Fourier transform, we convert the time domain data into its corresponding frequency-domain representation, enabling the extraction of intricate details about the material properties. This process is crucial for understanding how the material behaves under different frequency conditions, providing valuable insights that are not readily apparent in the time domain. The resulting CSM in the frequency domain provides a comprehensive view of the material's viscoelastic properties, facilitating more accurate simulations and predictions of its behavior under various conditions. According to Equation (9.13), it is evident that the CSM can be calculated directly in the following manner:

$$\varrho(x) = \Xi\left\{\frac{-\rho\omega_0^2\Omega_z(x,\omega_0)}{\nabla^2\Omega_z(x,\omega_0)}\right\} \tag{9.14}$$

$$\varsigma(x) = \Theta\left\{\frac{-\rho\omega_0^2\Omega_z\left(x,\omega_0\right)}{\nabla^2\Omega_z\left(x,\omega_0\right)}\right\} \tag{9.14}$$

in which $\Omega_z\left(x,\omega_0\right)$ is obtained by applying the Fourier transform at the given angular frequency ω_0, while $\nabla^2\Omega_z\left(x,\omega_0\right)$ is calculated with the discrete Laplace operator.

Algorithm 1 provides a comprehensive step-by-step guide on how to estimate the CSM.

Algorithm 1: Proposed CSM estimation algorithm.

1. Initialize the simulation.
2. Set the excitation frequency $f_0 = 150Hz$.
3. Produce a shear wave through needle oscillation.
4. Gather velocity data with noise at 120×120 spatial coordinates.
5. Apply a BPF to remove noise in the frequency domain.
6. Remove the filtered signal's transient components.
7. Apply a Fast Fourier Transform (FFT) to the signal that has been filtered.
8. Calculate the CSM at each spatial location using Equation (9.14).
9. Use a median filter to remove noise in the spatial domain.
10. Evaluate the performance of the estimated CSM.

End.

The proposed methodology estimates the CSM through a meticulously structured sequence of simulation and noise mitigation steps. Initially, the algorithm commences with the simulation setup (Step 1), defining the essential parameters required to simulate shear wave propagation within biological tissues. The excitation frequency is precisely set to 150 Hz (Step 2), which corresponds to the oscillation frequency of the needle used to generate shear waves within the tissue (Step 3). The velocities of these shear waves are then meticulously recorded at each spatial point within a 120×120 grid (Step 4). However, the collected velocity data is inherently noisy due to environmental interferences during the measurement process. To address this, the algorithm applies a BPF (Step 5), which effectively isolates the relevant frequency components around the set excitation frequency while discarding the extraneous high- and low-frequency noise. Following the BPF, the transient components of the signal are removed (Step 6) to enhance the stability and reliability of the data. Subsequently, a Fast Fourier Transform (FFT) is executed on the filtered signal (Step 7) to transition the data into the frequency domain, facilitating further analysis. The next step involves calculating the CSM at each spatial location utilizing Equation (14) (Step 8), thereby generating an initial image that maps the tissue's elasticity and viscosity properties. Recognizing that the CSM data may still be affected by spatial noise, a median filter is employed in the spatial domain (Step 9). This median filter effectively attenuates localized noise while preserving critical edge details for accurate tissue characterization. Finally, the algorithm evaluates the performance of the CSM estimation (Step 10) by comparing normalized errors and assessing the contrast and delineation capabilities between different tissue regions. This comprehensive process culminates in the production of a high-fidelity CSM image, ideal for noninvasive biomedical diagnostics. This innovative approach demonstrates significant improvements in noise reduction and image clarity, making it a valuable tool in the field of shear wave elastography and advancing the capabilities of real-time medical imaging systems.

9.3 NUMERICAL SIMULATION RESULTS

The simulation setup for this study involves a two-dimensional medium, measuring 120×120 units, which includes a tumor positioned at the coordinates (40 mm, 40 mm) with a radius of 20 mm. The elasticity of the surrounding medium is set to $\mu_1 = 6000$ Pa, and its viscosity to $\eta_1 = 1.2$ Pa.s. In contrast, the tumor's elasticity is $\mu_2 = 9000$ Pa, with a viscosity of $\eta_2 = 1.8$ Pa.s. The needle, which induces the shear waves, oscillates at a frequency of 150 Hz with an amplitude of 5 mm. The medium's bulk density is $\rho = 1000$ kg/m^3. The ideal initial image of the setup is depicted in Figure 9.2.

To elaborate further, the choice of parameters for the simulation is critical to accurately model the mechanical properties of biological tissues. The elasticity values (μ_1 and μ_2) represent the stiffness of the medium and the tumor, respectively, indicating how resistant these regions are to deformation. Higher elasticity in the tumor suggests it is stiffer compared to the surrounding tissue, a characteristic often observed in malignant tumors. Similarly, the viscosity values (η_1 and η_2) reflect the internal friction within the medium and the tumor, respectively. The higher viscosity in the tumor illustrates greater resistance to shear flow, which is another distinguishing feature when studying tissue mechanics. The oscillation frequency of the needle, set at 150 Hz, is carefully chosen to ensure effective shear wave propagation through the medium. This frequency is within the range typically used in shear wave elastography,

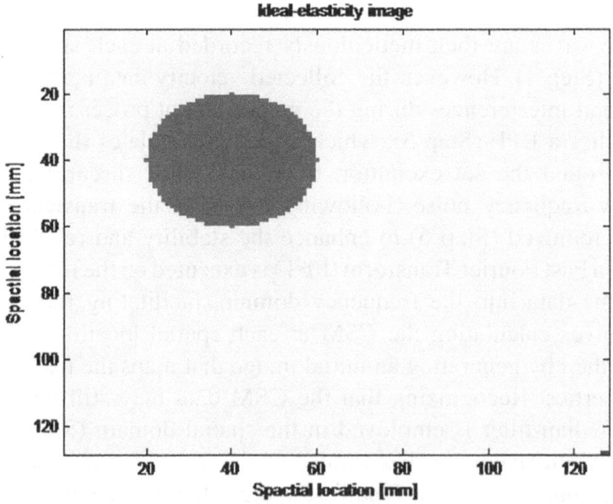

FIGURE 9.2 Ideal initial image.

as it balances the need for adequate wave penetration and resolution. The amplitude of 5 mm ensures sufficient displacement of tissue particles, facilitating the measurement of shear wave velocities. Additionally, the medium's bulk density (ρ = 1000 kg/m3) is set to mimic the density of biological tissues, providing a realistic simulation environment. The initial image in Figure 9.2 serves as a reference for evaluating the performance of the proposed algorithm in accurately estimating the CSM. This ideal image illustrates the expected distribution of elasticity and viscosity within the medium, against which the reconstructed images can be compared to assess accuracy and fidelity. By simulating such detailed and parameter-specific conditions, this study aims to enhance the understanding and application of shear wave elastography in biomedical imaging, particularly for tumor characterization and noninvasive diagnosis. The comprehensive setup ensures that the simulated data closely resembles actual tissue behavior, thereby providing a robust foundation for validating the proposed algorithm's effectiveness in real-world scenarios.

The reconstruction quality is very good when the spatial points are close to the vibrating needle, as demonstrated in Figure 9.3(a)–(d), 9.4(a)–(d), 9.5(a)–(d), and 9.6(a)–(d). Accurate estimations are provided by the region inside a quarter circle centered on the needle and having a radius of 120×120 mm. Beyond this region, however, the accuracy of the estimations decreases due to the rapid decline in particle velocity. Visually, it is apparent that our dual-filter method provides a more accurate estimation compared to the traditional one-filter approach, as it allows for a clearer observation of the object (with adequate contrast in elasticity and viscosity) within the area of interest, while with the traditional one-filter method, the recovered object image is shown quite faintly, and we cannot even observe the object in the case of viscous recovery. The ideal line is displayed by the gray line in Figure 9.3(e)–(l), 9.4(e)–(l), 9.5(e)–(l), and 9.6(e)–(l), while the estimated lines of elasticity and viscosity at horizontal lines 10 and 58 in the region of interest are displayed by the Dark Gray line. The results show that with the proposed dual-filter method, the estimated line follows the ideal line quite closely, especially at spatial positions near the excitation source, while at positions distant from the excitation source, the estimation is slightly worse, which is shown by small fluctuations. Meanwhile, with the traditional one-filter method, we see that the fluctuations appear quite a lot and are large, especially at the boundary positions between the background environment and the object, and spatial points far from the source of excitation.

The normalized errors for elasticity and viscosity estimation at various time steps are presented in Table 9.1. For time steps of 500, 1000, 1500, and 2000, the proposed dual-filter method achieves reductions in normalized error for elasticity estimation by 4.54%, 19.49%, 15.68%, and 16.95%, respectively, compared to the traditional one-filter method. Likewise, for the same time steps, the reductions in normalized error for viscosity estimation using the proposed dual-filter method are 12.70%, 31.87%, 43.70%, and 53.04%, respectively. These findings indicate a marked enhancement in image quality for both elasticity and viscosity estimations choosing the suggested strategy over the traditional one.

This significant reduction in error highlights the efficacy of the suggested dual-filter approach, especially the application of a median filter in conjunction with a BPF. The BPF initially reduces unwanted frequency components, ensuring the collected data

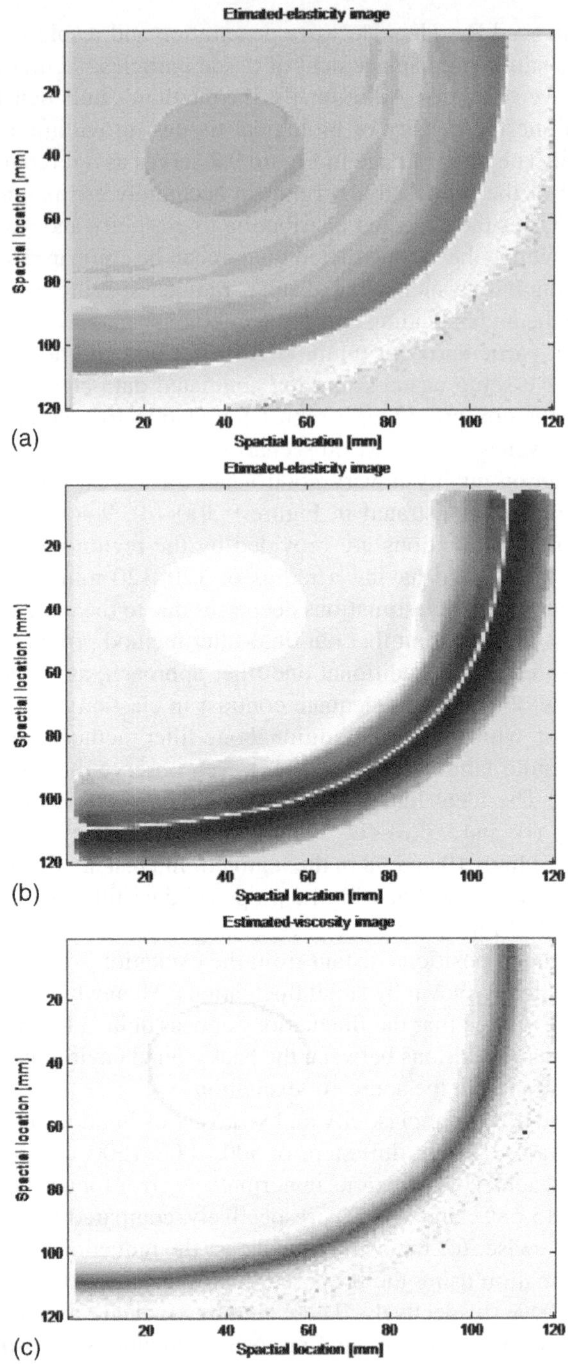

FIGURE 9.3 Reconstruction results for elasticity and viscosity at spatial points close to the vibrating needle. (a)–(d) Reconstructed parameter maps demonstrating high reconstruction quality in regions proximal to the excitation source. (e)–(l) Comparison between the ideal reference line (gray) and the estimated elasticity and viscosity profiles (dark gray) along horizontal lines 10 and 58 within the region of interest.

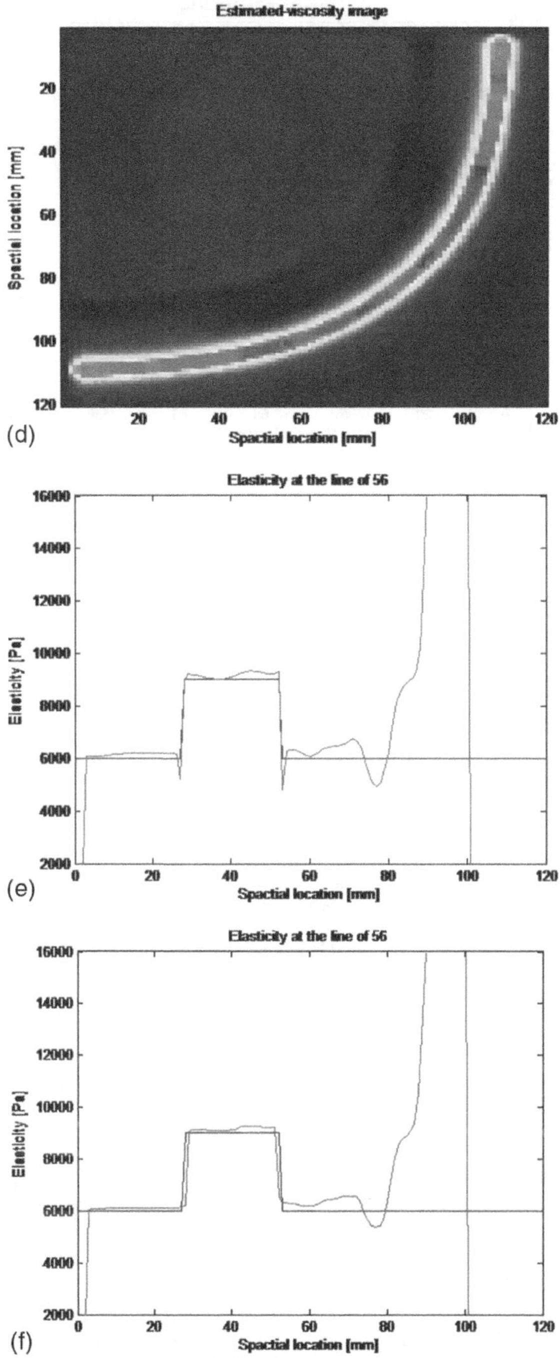

FIGURE 9.3 (Continued)

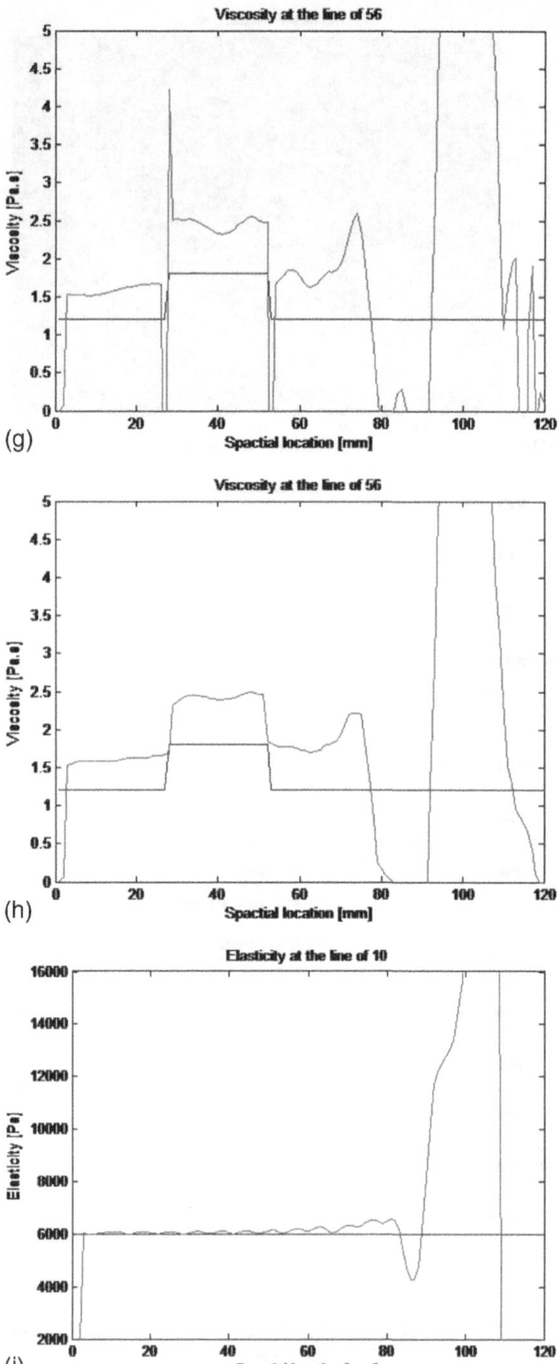

FIGURE 9.3 (Continued)

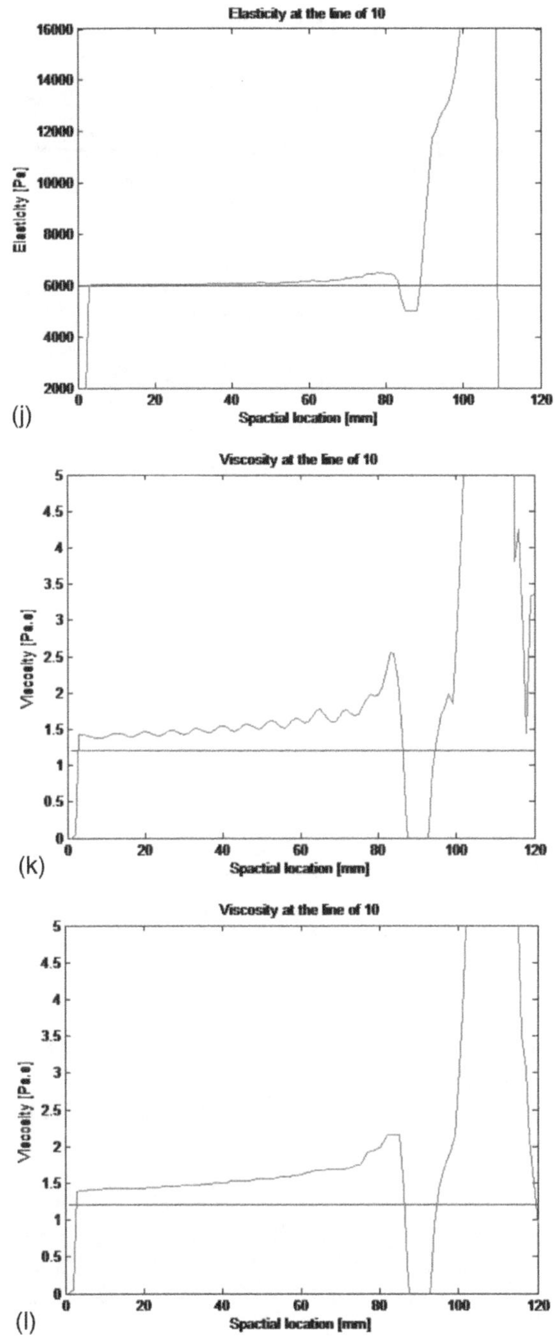

FIGURE 9.3 (Continued)

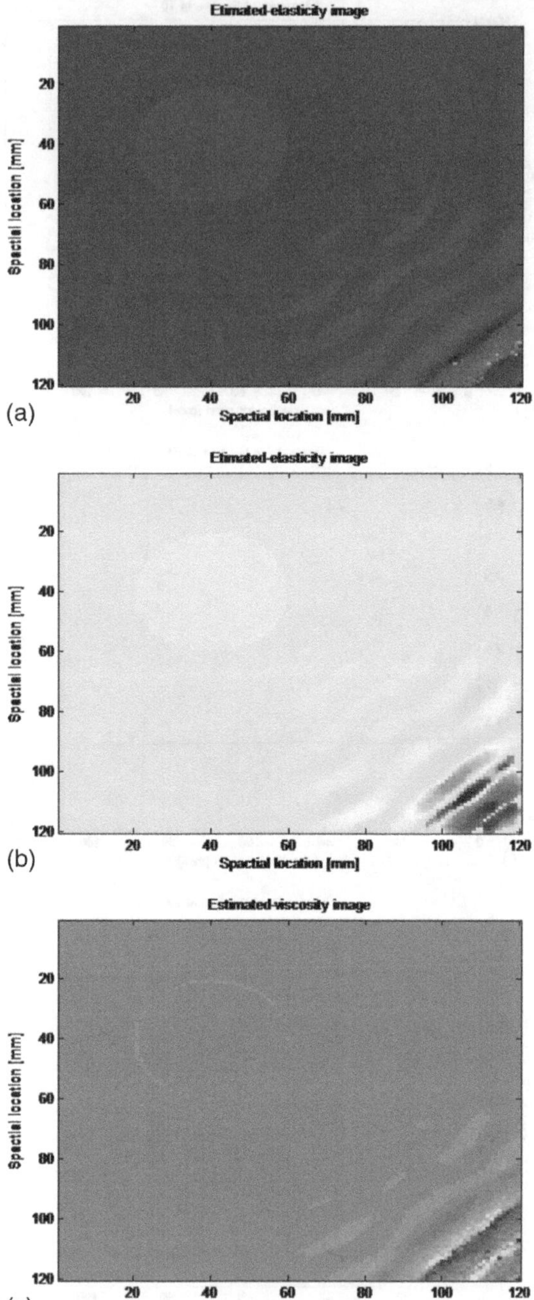

FIGURE 9.4 Reconstruction performance at intermediate spatial locations relative to the vibrating needle. (a)–(d) Parameter reconstruction results showing the effect of increased distance from the excitation source. (e)–(l) Ideal reference line (gray) and estimated elasticity and viscosity profiles (dark gray) extracted along horizontal lines 10 and 58 of the region of interest.

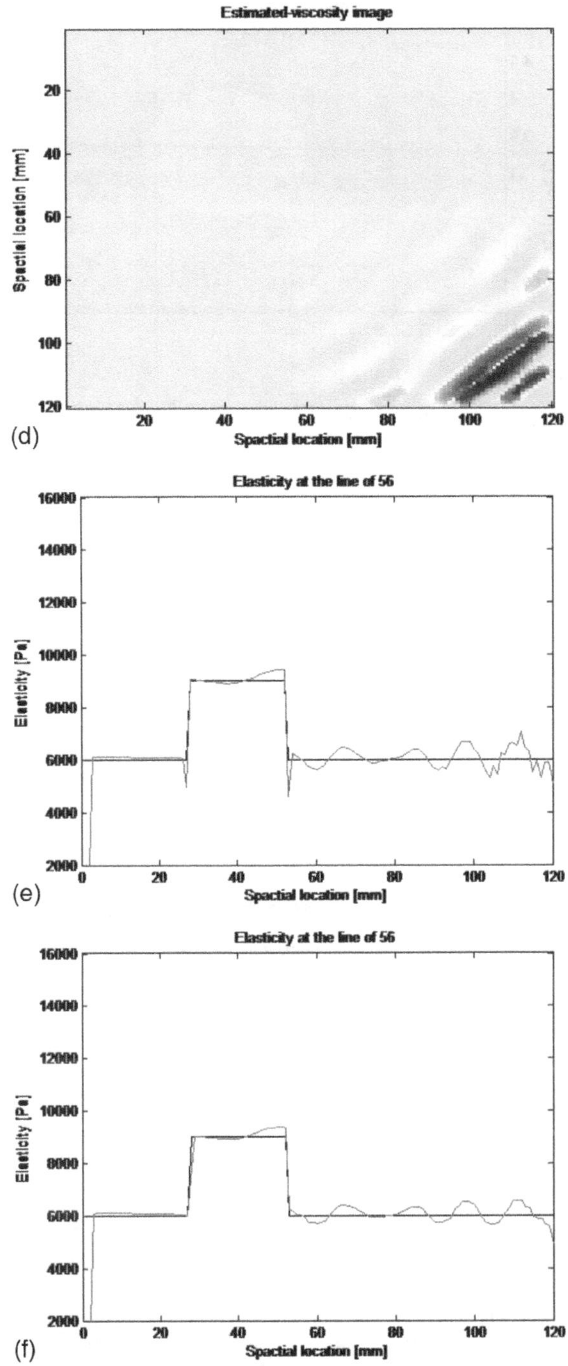

FIGURE 9.4 (Continued)

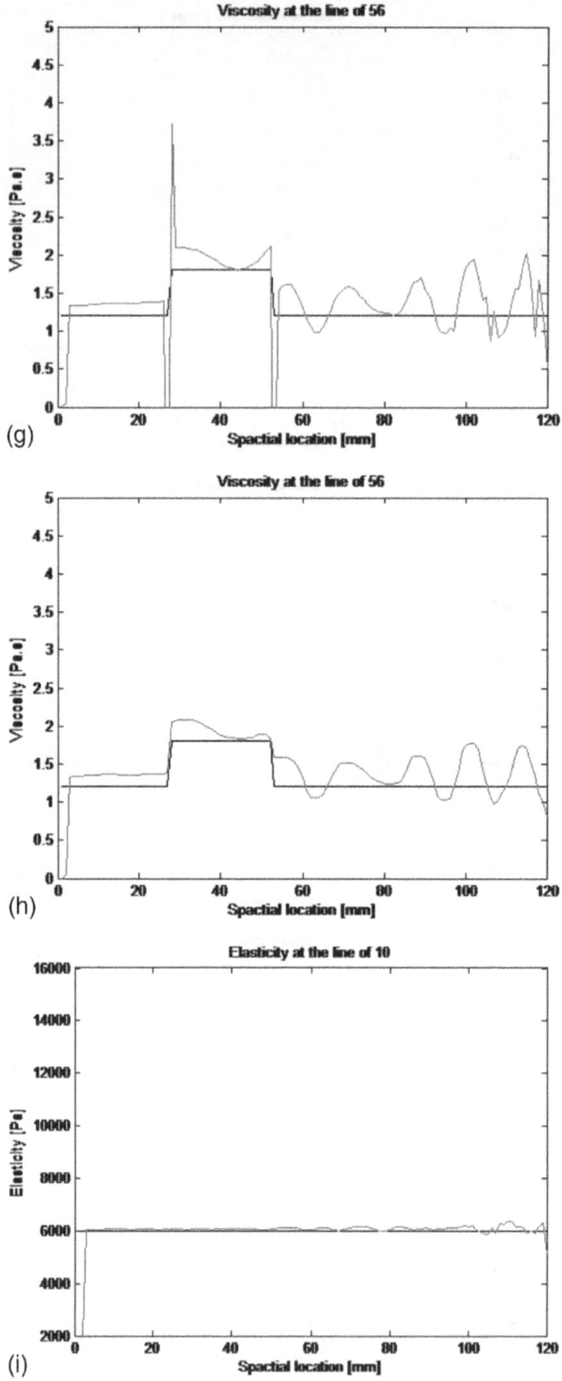

FIGURE 9.4 (Continued)

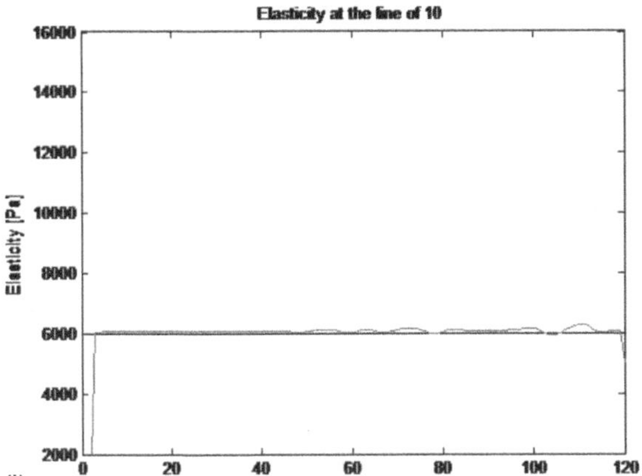

(j)

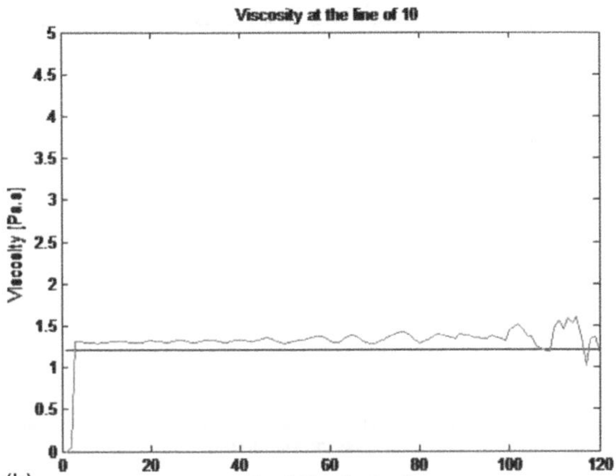

(k)

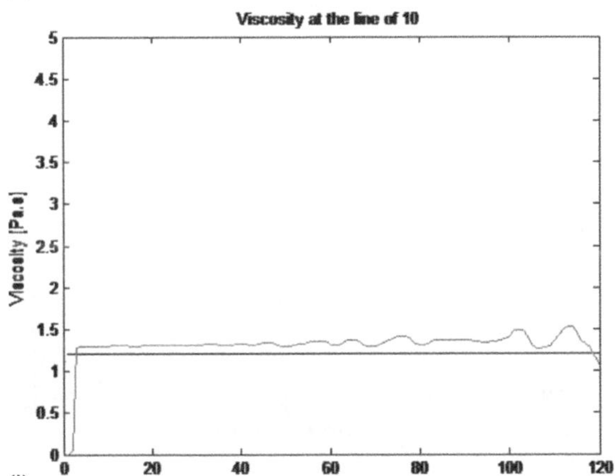

(l)

FIGURE 9.4 (Continued)

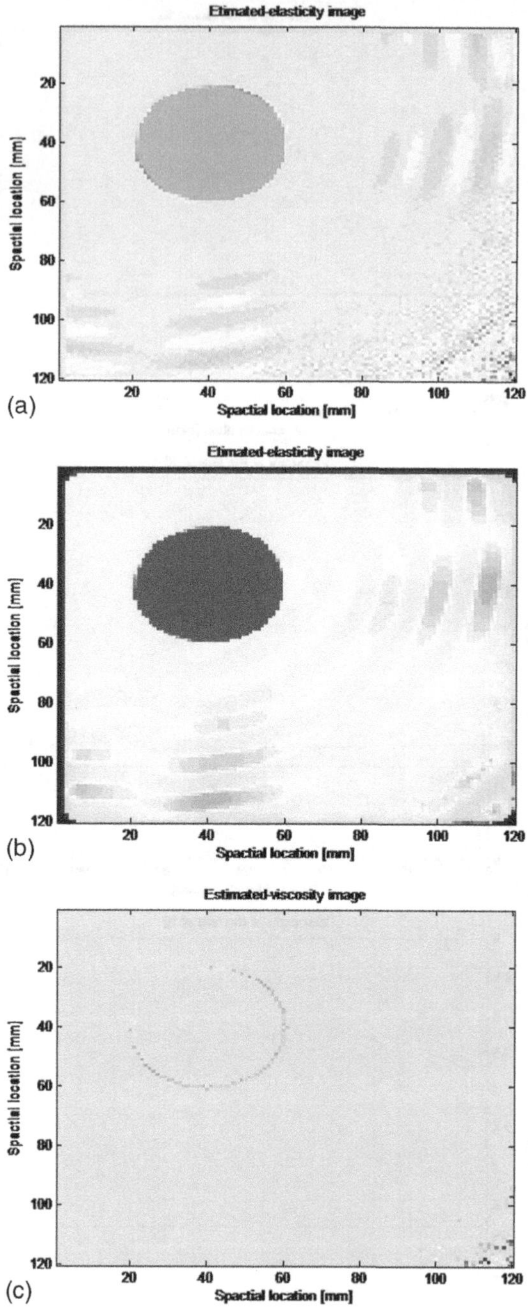

FIGURE 9.5 Elasticity and viscosity reconstruction under varying spatial conditions (a)–(d) Reconstructed elasticity and viscosity distributions at selected spatial points. (e)–(l) Line-profile comparison of the ideal reference (gray) with estimated elasticity and viscosity (dark gray) at horizontal lines 10 and 58 in the region of interest.

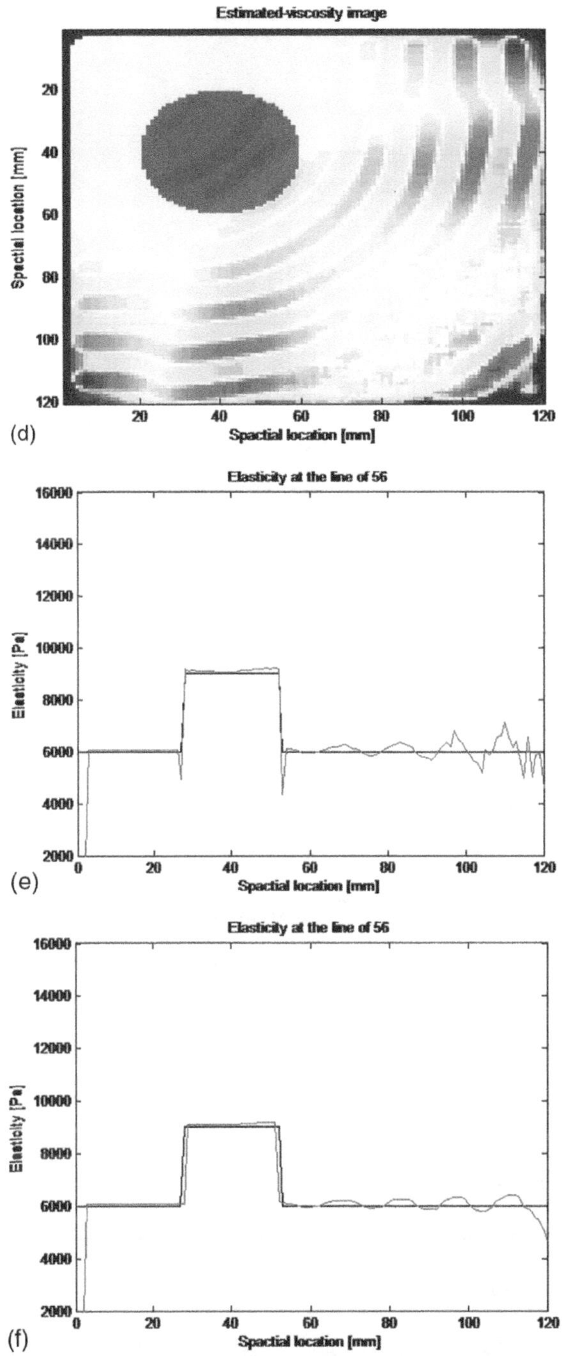

FIGURE 9.5 (Continued)

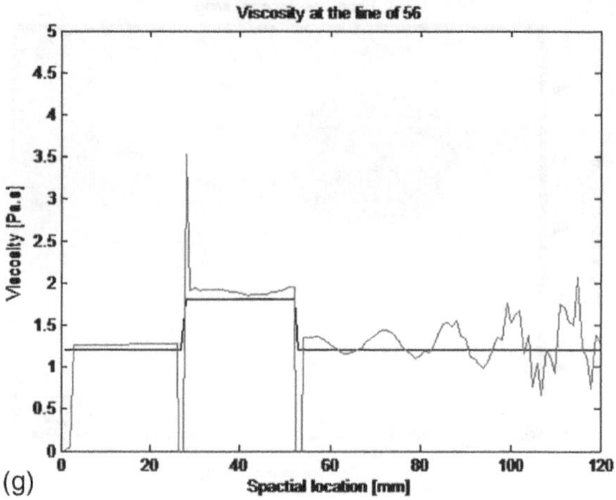

(g)

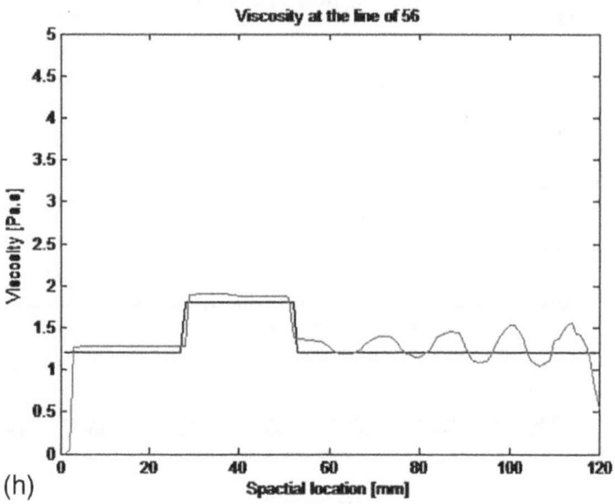

(h)

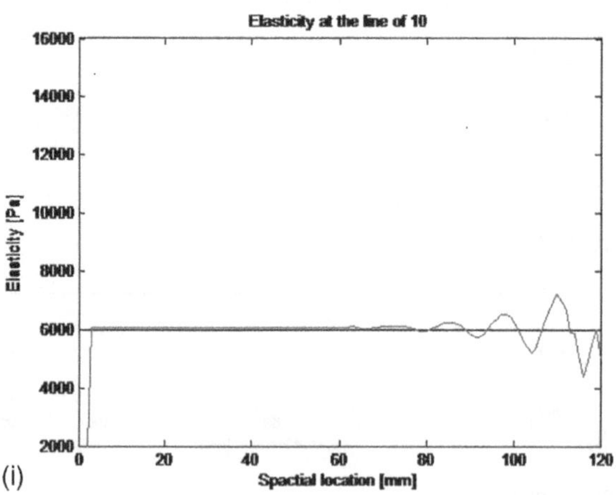

(i)

FIGURE 9.5 (Continued)

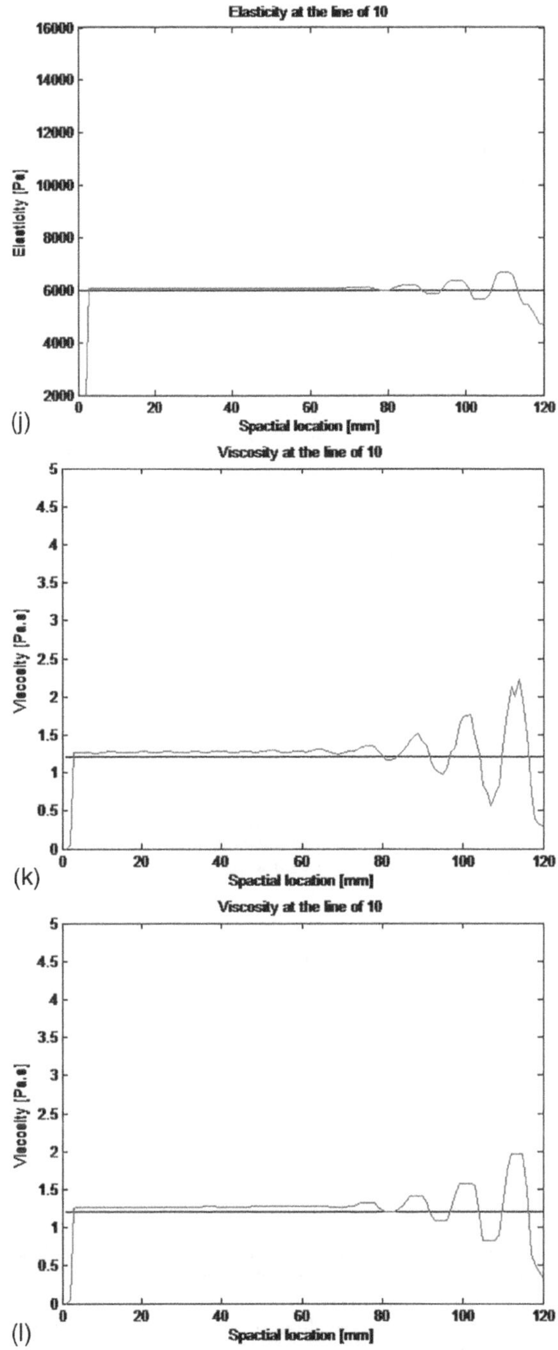

FIGURE 9.5 (Continued)

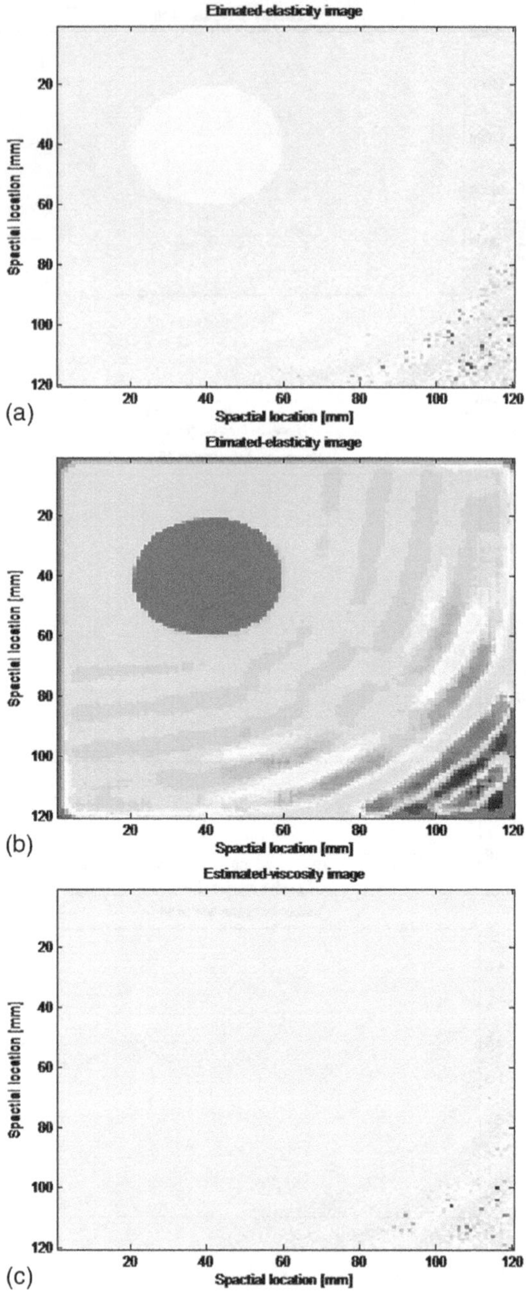

FIGURE 9.6 Reconstruction accuracy evaluation at farther distances from the vibrating needle. (a)–(d) Reconstructed elasticity and viscosity maps illustrating degradation in reconstruction quality with increasing distance. (e)–(l) Comparison of the ideal line (gray) and estimated elasticity and viscosity profiles (dark gray) along horizontal lines 10 and 58 in the region of interest.

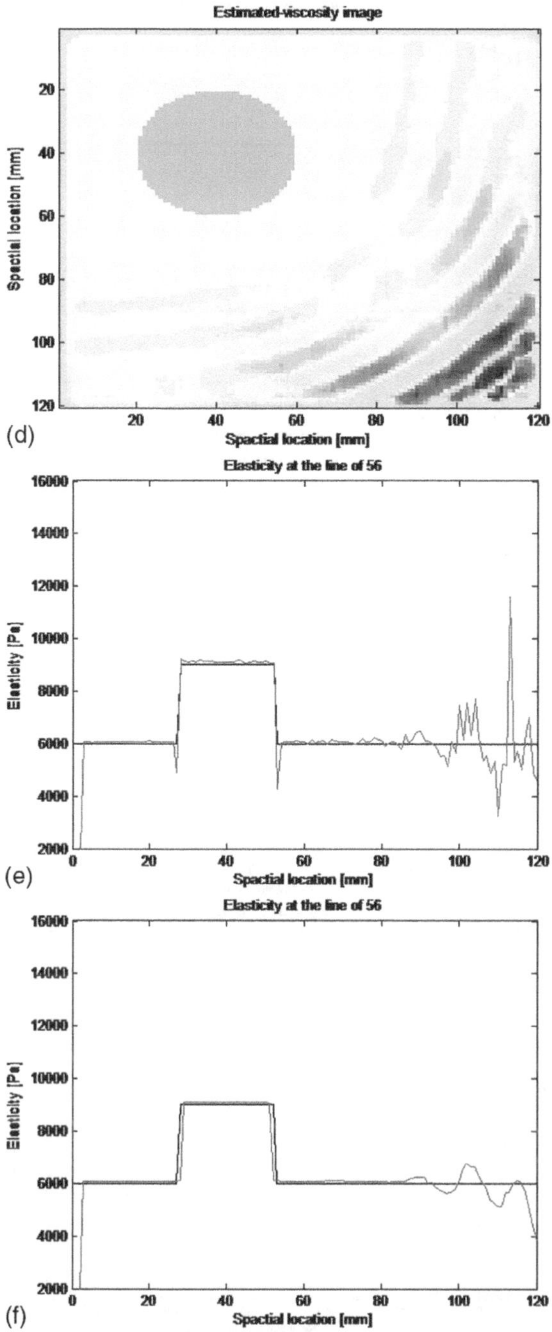

FIGURE 9.6 (Continued)

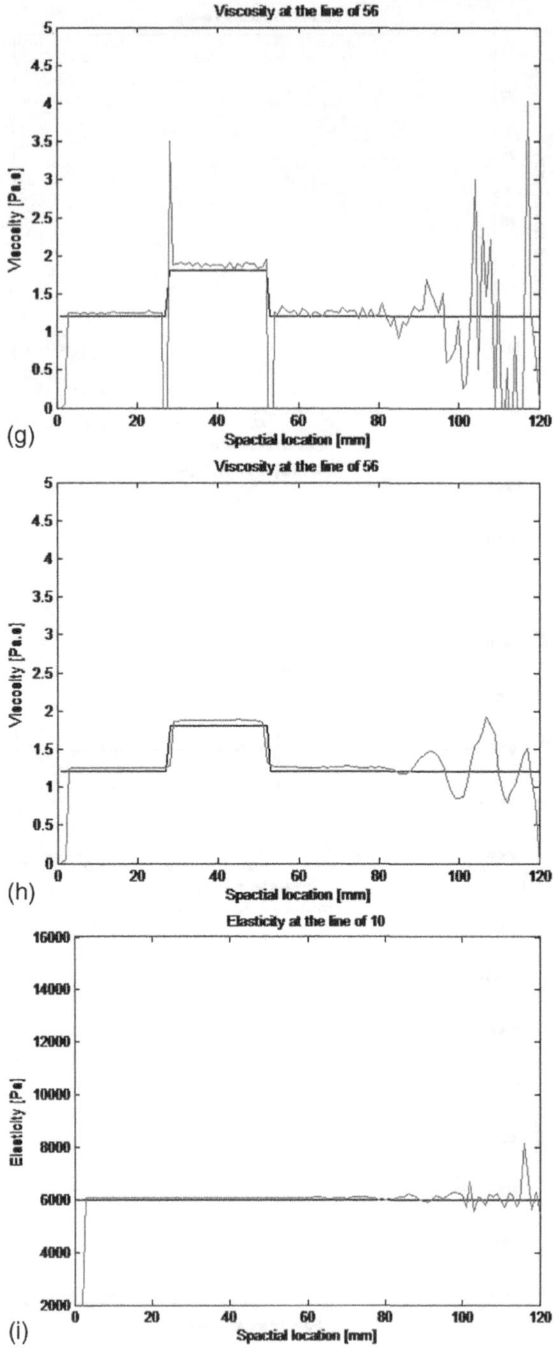

FIGURE 9.6 (Continued)

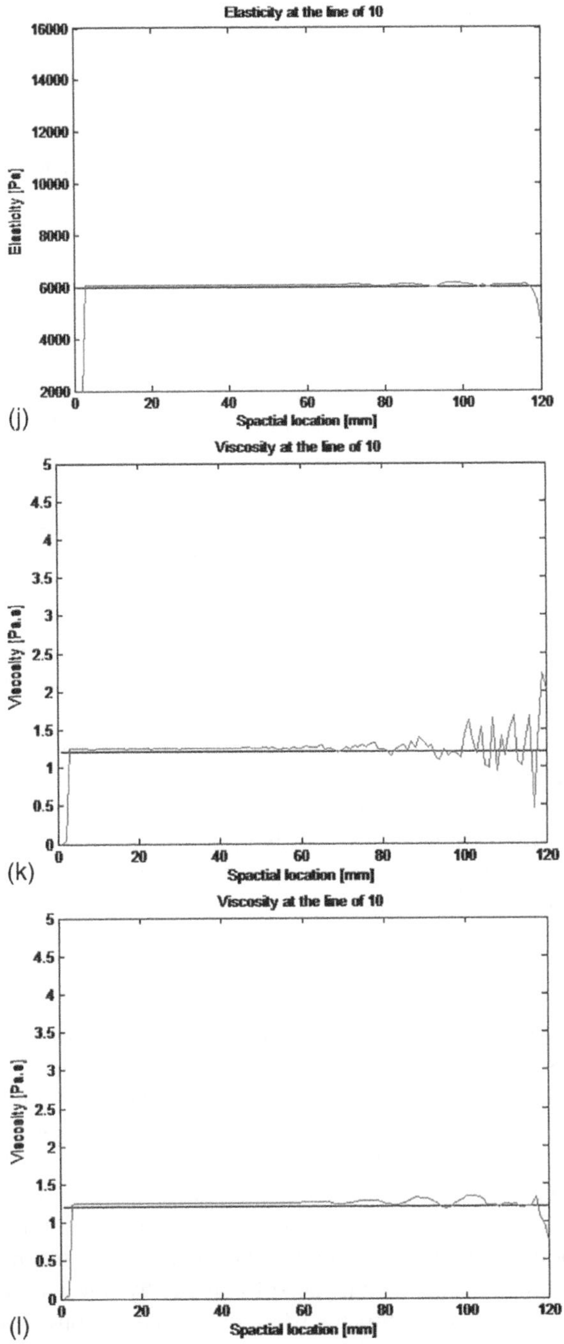

FIGURE 9.6 (Continued)

TABLE 9.1
Normalized Error in the Estimated Techniques for Elasticity and Viscosity across Varying Time Steps

Methods	Time steps	500	1000	1500	2000
The conventional method	Estimated elasticity	0.0617	0.0236	0.0185	0.0177
	Estimated viscosity	0.4434	0.1735	0.1192	0.0988
The proposed method	Estimated elasticity	0.0589	0.0190	0.0156	0.0147
	Estimated viscosity	0.3871	0.1182	0.0671	0.0464

focuses on the frequency band surrounding the excitation frequency. This step is crucial for minimizing noise in the initial stages of data processing. However, residual spatial noise—such as random fluctuations and localized disturbances—can still affect the accuracy of CSM estimates. Here, the median filter proves particularly advantageous.

Applying a median filter in the spatial domain effectively removes this residual noise without compromising the clarity of boundaries and edges, which are essential for distinguishing tissue regions with different mechanical properties. Unlike mean or Gaussian filters, which may blur critical edges, the median filter preserves important structural details, enhancing contrast between adjacent tissue regions. This property is essential for biomedical imaging applications, where precise boundary information can be vital for diagnostics.

Overall, the combined use of a BPF in the frequency domain and a median filter in the spatial domain not only achieves greater accuracy by significantly lowering error rates but also provides a robust approach that retains essential details in reconstructed images. This two-filter strategy enhances both the fidelity and practical applicability of CSM imaging, making it a valuable solution for real-time medical imaging systems where accuracy and clarity are critical.

9.4 CONCLUSIONS

This chapter introduces an advanced methodology for reconstructing CSM images of biological tissues by integrating frequency-domain BPF with spatial-domain median filtering. The proposed method employs the FDTD model to simulate shear wave propagation through heterogeneous tissues. Subsequently, a bandpass filter (BPF) is used to suppress both high- and low-frequency noise surrounding the excitation frequency. Following direct CSM estimation using the AHI algorithm, a median filter is applied to the resulting image to reduce spatial noise while preserving critical boundary details. Simulation results demonstrate that this dual-filtering technique significantly enhances image quality. This is evidenced by markedly lower normalized errors in both elasticity and viscosity estimates when compared to conventional single-filter methods. The technique's ability to provide improved contrast and detail retention between different tissue regions highlights its potential for accurate, noninvasive diagnostics. Moreover, the dual-filtering strategy minimizes computational complexity, enabling the feasibility of real-time CSM imaging without the need for adaptive or optimal-filtering techniques. Overall, this innovative approach reveals a promising pathway toward practical, high-fidelity viscoelasticity imaging. The enhanced image clarity and precision indicate its potential contribution to advancements in diagnostic imaging technologies for healthcare applications, offering substantial improvements over traditional methods. By providing more reliable and detailed images, this method could greatly aid in the early detection and diagnosis of various medical conditions, thereby improving patient outcomes and advancing the field of biomedical imaging.

REFERENCES

[1] Garra, B. S. (2007). Imaging and estimation of tissue elasticity by ultrasound. *Ultrasound Quarterly*, 23(4), 255–268.

[2] Sigrist, R. M. S., Liau, J., Kaffas, A. E., Chammas, M. C., & Willmann, J. K. (2017). Ultrasound elastography: Review of techniques and clinical applications. *Theranostics*, 7(5), 1303–1329.

[3] Chen, S., Urban, M. W., Pislaru, C., Kinnick, R., Zheng, Y., Yao, A., & Greenleaf, J. F. (2009). Shearwave dispersion ultrasound vibrometry (SDUV) for measuring tissue elasticity and viscosity. *IEEE Transactions on Ultrasonics, Ferroelectrics, and Frequency Control*, 56(1), 55–62.

[4] Parker, K. J., Doyley, M. M., & Rubens, D. J. (2011). Imaging the elastic properties of tissue: The 20 year perspective. *Physics in Medicine & Biology*, 56(1), R1–R29.

[5] Jha, S., Seo, C., Yang, E., & Joshi, G.P. (2021). Real time object detection and trackingsystem for video surveillance system. *Multimediea Tools and Applications*, 80(3), 3981–3996. https://doi.org/10.1007/s11042-020-09749-x.

[6] Nightingale, K. (2011). Acoustic radiation force impulse (ARFI) imaging: A review. *Current Medical Imaging Reviews*, 7(4), 328–339.

[7] Catheline, S., Gennisson, J. L., Delon, G., Fink, M., Sinkus, R., Abouelkaram, S., & Culioli, J. (2004). Measurement of viscoelastic properties of homogeneous soft solid using transient elastography: An inverse problem approach. *The Journal of the Acoustical Society of America*, 116(6), 3734–3741.

[8] Palmeri, M. L., & Nightingale, K. R. (2011). Acoustic radiation force-based elasticity imaging methods. *Interface Focus*, 1(4), 553–564.

[9] Sarvazyan, A. P., Rudenko, O. V., Swanson, S. D. , Fowlkes, J. B., & Emelianov, S. Y. (1998). Shear wave elasticity imaging: A new ultrasonic technology of medical diagnostics. *Ultrasound in Medicine & Biology*, 24(9), 1419–1435.

[10] Jha, S., Prashar, D., Long, H. V., & Taniar, D. (2020). Recurrent neural network for detecting malware. *Computers & Security*, 99, 102037.

[11] Marko, O., & Insana, M. F. (2010). Model-based complex shear modulus reconstruction: A Bayesian approach. *IEEE Int'l Ultrasonics Symposium*, pp. 61–64. IEEE Press.

[12] Jha, S., Nkenyereye, L., Joshi, G. P., & Yang, E. (2020). Mitigating and monitoring smart city using internet of things. *Computers, Materials & Continua*, 65(2), 1059–1079.

[13] Chen, S., Urban, M. W., Pislaru, C., Kinnick, R., Zheng, Y., Yao, A., & Greenleaf, J. F. (2009). Shearwave dispersion ultrasound vibrometry (SDUV) for measuring tissue elasticity and viscosity. *IEEE Transactions on Ultrasonics, Ferroelectrics, and Frequency Control*, 56(1), 55–62.

[14] Nasief, H., Rosado-Mendez, I., Zagzebki, J. A., & Hall, T. (2016). Detecting anisotropy in attenuation coefficient of in-vivo human breast tumors. *2016 WARF Discovery Challenge*, 2016.

[15] Coila, A., Rouyer, J., Zenteno, O., & Lavarello, R. (2016). A regularization approach for ultrasonic attenuation imaging. *2016 IEEE International Symposium on Biomedical Imaging*, Prague, pp. 469–472.

[16] Hou, S. W., Merkle, A. N., Babb, J. S., McCabe, R., Gyftopoulos, S., & Adler, R. S. (2017). Shear wave ultrasound elastographic evaluation of the rotator Cuff Tendon. *Journal of Ultrasound in Medicine*, 36(1), 95–106.

[17] Jha, S., Prashar, D., & Elngar, A. A. (2020). A novel approach using Modified Filtering Algorithm (MFA) for effective completion of cloud tasks. *Journal of Intelligent & Fuzzy Systems*, 39, 8409–8417.

[18] Orescanin, M., Wang, Y., & Insana, M. F. (2011). 3D fdtd simulation of shear waves for evaluation of complex modulus imaging. *IEEE Transactions on Ultrasonics, Ferroelectrics and Frequency Control*, 58, 389–398.

[19] Luong Quang, H., Nguyen, T. T., & Vo, V. N. (2018). Complex shear modulus estimation using integration of LMS/AHI algorithm. *International Journal of Advanced Computer Science and Applications (IJACSA)*, 9(8), 584–589.

[20] Orescanin, M., & Insana, M. F. (2010). Model-based complex shear modulus reconstruction: A Bayesian approach. *Ultrasonics Symposium*, IEEE, pp. 61–64.

[21] Orescanin, M., Wang, Y., & Insana, M. F. (2011). 3d fdtd simulation of shear waves for evaluation of complex modulus imaging. *IEEE Transactions on Ultrasonics, Ferroelectrics, and Frequency Control*, 58(2), 389–398.

[22] Papazoglou, S., Hamhaber, U., Braun, J., & Sack, I. (2008). Algebraic helmholtz inversion in planar magnetic resonance elastography. *Physics in Medicine and Biology*, 53(12), 3147.

10 A Two-Stage Auction Mechanism for Cloud Resource Allocation for Artificial Intelligence and Machine Learning

Salliah Shafi, Sultan Ahmad, Gufran Ahmad Ansari, and Sudan Jha

10.1 INTRODUCTION

In modern society, authors collect a lot of data from different sources like customer and cloud computing is a platform. These companies provide various services to handle and process data. As demand for these services grows, companies are starting to share their extra resources with each other to improve services they offer to customers. This means that one company can temporarily rent spare resources from another company. In the future, large companies are expected to be able to control the cloud business by hiring resources from small companies. To decide how resources are allocated in this situation, an auction mechanism can be used. An auction is a simple way to determine who gets what resources. Recently, Amazon introduced a framework called Spot Instances, which is an auction-based system for resource allocation. In this system, users bid for the cloud servers they need. The resources are then allocated based on the bids submitted by users. The auction mechanism is a good choice for modeling resource allocation because it is simple, versatile, and matches well with the way cloud networks work [1, 2]. It allows users to bid for resources based on how much they are willing to pay. This can potentially save money compared to fixed pricing models. To summarize, cloud computing is important for processing large amounts of data, and companies like Microsoft, Google, and Amazon provide cloud services. Sharing resources between companies can improve services, and auctions like Amazon Spot Instances can be used to allocate resources. Auctions are simple and effective for determining who gets the resources they need. In today's cloud network, there are many types of cloud servers with different hardware and software configurations [3–5]. These servers are used to perform various tasks and meet the needs of different users. Sometimes a combination of different server types is needed to meet the different needs of users. Instance can switch between busy and idle, customers can join or leave cloud business at any time. To deal with the

DOI: 10.1201/9781003609773-10

change it is better to have multiple contests called multiple contests rather than a single contest. The problem is that an auction used in a competitive auction where people bid on servers will not do well in sequential auctions. In a onetime auction, customers are encouraged to bid honestly because they think the server is right for them. This is called real estate. But as the series of auctions continue, it will get harder and harder to get the truth [6–9]. This means that customers will try to control their bids for better result which is not fair to other contributors. To meet this challenge, we must find new ways to create a competitive system that works well in regular competition while also encouraging participants to be honest during the competition. This ensures fairness and prevents manipulation that could cause some customers to be dishonest. In summary, today's cloud network requires different types of servers, and competitive auctions are better for checking the quality of the cloud server and the cooperation of customers. However, it is difficult to ensure the accuracy of these competitions, so we must develop new competitions that are both regular and fair.

10.2 LITERATURE SURVEY

In this section alternatives-primarily based sequential auctions for dynamic cloud useful resource allocation. With growing desires for cloud computing services, the concept of managing constrained cloud resources for growing an earnings has arisen as a crucial problem. Auction idea is presently considered as a viable way to solve the problem of cloud useful resource allocation. On this chapter, remember a model for Cloud of Clouds Networks (CCNs) with top-notch kinds of servers alongside clients with heterogeneous desires, wherein clients and cloud servers also can moreover be a part of and leave the CCN at will. We endorse an options-based totally sequential public sale that not handiest gives an terrific healthful with the dynamic shape of the trouble, however additionally solves the doorway time hassle and possesses the truthfulness assets and observe every first-rate and second-charge alternatives-primarily based totally sequential auctions, and model the fee matching methods inside the ones auctions as Markov chains P Wang and E Hossain [10] Auction-based Resource Allocation in Cognitive Radio (CR) Systems Auction concepts as a subfield of economics provide useful tools for modeling, analysis, and optimization of radio systems for effective management in the radiological environment. Through the use of open sales, radio equipment, including broadcast channels, time and quality, can be shared to the system by certified and unauthorized users according to business rules. Due to the capability of the docking mechanism, there are many broadcasting methods implemented in private radio stations. This chapter first explains why auctions are appreciated. The motivation and main problem of using auctions for Wi-Fi network architectures and protocols are discussed Using Spectrum Heterogeneity in the Dynamic Spectrum Market (DSM) is an important wireless tool that provides opportunistic spectrum access to alleviate the spectrum scarcity problem. DSM allows traditional spectrum owners to lease their channels to unlicensed spectrum users (or secondary users) as a way to increase revenue and improve spectrum usage. At DSM, the cost of renting more spectrum for Wi-Fi wireless is an important but smart issue that requires a good understanding of partner interests and interactions copy.

This chapter publishers (WSPs) lease access to spectrum and secondary users (SUs) Purchase spectrum usage to maximize their usage. Selected important but previously excluded DSM buildings: 1) heterogeneous spectrum assets of WSPs and 2) spectrum sharing among SUs. Analysis of the effect of spectral heterogeneity from long-term use See software usage – Wi-Fi radio (SDR) See bands. A three-step process is used to see the impact of spectral heterogeneity on premium rates for WSPs and WSP strategy expectations for SUs. First, we test how spectrum sharing among SUs registered to the same WSP affects the effectiveness of SUs. Then, given the difference between variables and costs, the WSP method's maximum expectation for SU is given to obtain the best value. Y Xu and X Wang [11] Multi-Auction Tiered Auction CR for Dynamic Spectrum Access (DSA) is a promising approach to access limited resources by allowing users to license access to spectrum used Green, and DSA is one of them. Resources of CR networks. Public sale of the spectrum to the market has been described as a powerful way to obtain DSA. In the Spectrum Public Sale, the Number 1 Spectrum Owner (PO) acts as the auctioneer, choosing to sell the credits for additional income, and the secondary user (SU) acting as the bidder, choosing to buy the spectrum bands IT or service. However, the traditional auction model is limited to the case of single auctioneers. This chapter examines a public-spectrum sale with several auctioneers and several bidders, which is better suited to smart CR networks. Proponents of MAP are the Multiple-Auctioneer Tier public sale mechanism, in which each auctioneer systematically raises the bid and ask price, and each bidder will sooner or later choose an auctioneer to bid on. Equilibrium explained that neither auctioneer nor bidder is willing to change their orders. Y Xu and Q Zhan [12] Spectrum Trading in Smart Radio: Solving Theoretical Modeling Market oriented spectrum trading is a good way to achieve dynamic spectrum acquisition/sharing. This chapter analyzes the difficulty of spectrum trading where the primary spectrum owner (or primary person, PO) sells their inert spectrum to more than one SU. Model the market as a monopoly market in which the PO, allocating features and prices for the spectrum it sells, acts as a monopoly and SU acts as the customer and selects the spectrum with the costs and the costs to be obtained. Monopoly-dominated premium-contract terms regulated by POs, including a combination of premium connections, all of which affect customers. Compromise can be reached if the incentive of each SU is similar (IC) and in my case the compromise (IR) buys spectrum with certain values suitable for its type. Recognize the importance and risks that make a contract possible. An agreed protocol is the same and optimizes and enables the software of the PO, which is a separate version for each customer and does not affect the receiver type model [13–15]. Additionally, by analyzing the balance of the relationship, i.e., the integration of each PO and SU, we see that the balance in society will be less than the most effective policy or close to the best results, depending on the distribution of customers. B Li and W Wang [16] Optimizing Auctions through Dynamic Auctions in the IaaS Cloud Market Cloud pricing plays an important role in the success of cloud computing. However, none of the current pricing plans provide a guarantee of service (like Spot Instances in Amazon EC2) or use on-demand pricing where prices are not responding to demand in the short-term time domain in Amazon EC2). To address these issues, this document has taken the dynamic auction form of auctions that are calculated periodically to meet buyers' needs throughout the year. Two important effects are discussed, namely sales maximization and sales accuracy. Our setup includes a resource allocation plan that determines the number of events to compete at one time, similar to a potential-auction, as most of the fees are similar to the allocation of resources planned over the years [17–20]. We show that our layout is real in two dimensions and is maximally fit asymptotically even if the name is large enough. Resources planned over the years. We show that our layout is real in two dimensions and is maximally fit asymptotically even if the name is large enough.

10.3 PROBLEM DEFINITION

The interactions among cloud providers (CPs) are modeled through a coalition game in which the CPs borrow belongings from every one of-a-kind to fulfil their client's needs. This work is most of the first to take into account the interactions every the various clients and cloud carriers, and a number of the CPs themselves; regrettably, clients with package desires had been now no longer taken into consideration. In addition, most of the tasks referred to do no longer remember the fact that one of the main demands for powerful air distribution is the hassle of sluggish get entry. This problem occurs when consumers delay entering the market with some marginal information about the future of the market. Assuming that in a competitive network, some users know that it can be obtained at low cost by using their cloud assets for the desired period of time. In this case, all customers with service information will delay their entry to the market. In large markets, this can cause spikes, and therefore market volatility.

10.4 EXISTING SYSTEM

Authors propose a realistic strategy for competitive bidding. In this framework, in addition to competitive decision and payment analysis, users' activities have to bid in a number of competitive bids when it is desired to take more than one unit of time for a cloud package. This fact renders the framework useless when users need to work on their tasks without interruption, and the interaction between customers and CPs is modeled as an online competition. The performance model captures the heterogeneity of different instance types and customer needs. It also addresses competition, where the decision is defined as one of the environmental optimization problems. They review actual sales and propose an approximation algorithm for the optimization environment. However, similarly, many optimization problems need to be solved using empirical methods at every difficulty, which complicates the framework and wastes a lot of money. All of the above work and most of the available documentation focus on modeling the interaction between the cloud manager and the client, while saving the process manager's resources. The cloud has been largely ignored.

10.5 PROPOSED SYSTEM

The system introduces a new system open framework to capture the interaction between (a) client and cloud management and (b) cloud management and CP. In particular, don't forget the cloud-to-cloud network (CCN) with different instances and clients. Each CCN has a CCN Manager who manages its assets. CCN managers are willing to hire CP servers to expand their resources, attract more customers and manage their real needs. The framework is powered by the contention-on-demand (OBSA) and simulates the interaction between the client and the CCN administrator; here all players get the resources they need from CCN. To our knowledge, we are the first to use OBSA (among other things) to address this important area of cloud communications. Similarly, we provide similar performance measures that are new to existing studies in the relevant literature. The second level of the framework defines the interaction between multiple CCN leaders and multiple CPs, where CCN leaders compete to gather resources from CPs. During this time, the special model was fully introduced, including the parallel market for the storage of aerial products: the parallel market and market-based between the client and the CCN administrator; here all players get the resources they need from CCN. To our knowledge, we are the first to use OBSA (among other things) to address this important area of cloud communications. Similarly, we provide similar performance similar performance measures that are new to existing studies in the relevant literature. The second level of the framework defines the interaction between multiple CCN leaders and multiple CPs; where CCN leaders compete to gather resources from CPs. During this time, the special model was fully introduced, including the parallel market for the storage of aerial products: the parallel market and market-based sales. For better Capture of the property.

10.6 ALGORITHM

Triple DES is a block cipher that applies the DES algorithm thrice. It generally makes use of three unique keys K1, K2, and K3. The primary key, K1, is used to encrypt, the second one key, K2, is used to decrypt, and the 1/3 key, K3, is used to encrypt once more. The triple DES additionally has a variation that makes use of only keys, in which K1 and K3 are the identical. Triple DES algorithm Steps let us check the stairs involved inside the DES set of rules:

The preliminary permutation (IP) feature gets the 64-bit plaintext block.

- IP is completed on plaintext.
- IP then makes halves of the block that has been permuted.
- Two halves are referred to as left plaintext (LPT) and proper text (RPT).
- All LPTs and RPTs are encrypted 16 instances

10.7 SYSTEM ARCHITECTURE

The proposed system architecture is illustrated in Figure 10.1, showing the overall design and interaction between the IoT sensors, data acquisition module, processing unit, and user interface. Each component is interconnected to ensure seamless data collection, processing, and visualization. At the same time others are suitable for real-time database computation and parallel processing. In addition to its central servers, CCN can lease servers from CP to manage backup operations and support additional customers. CPs are small shops that lease larger assets to CCN for income. Customers

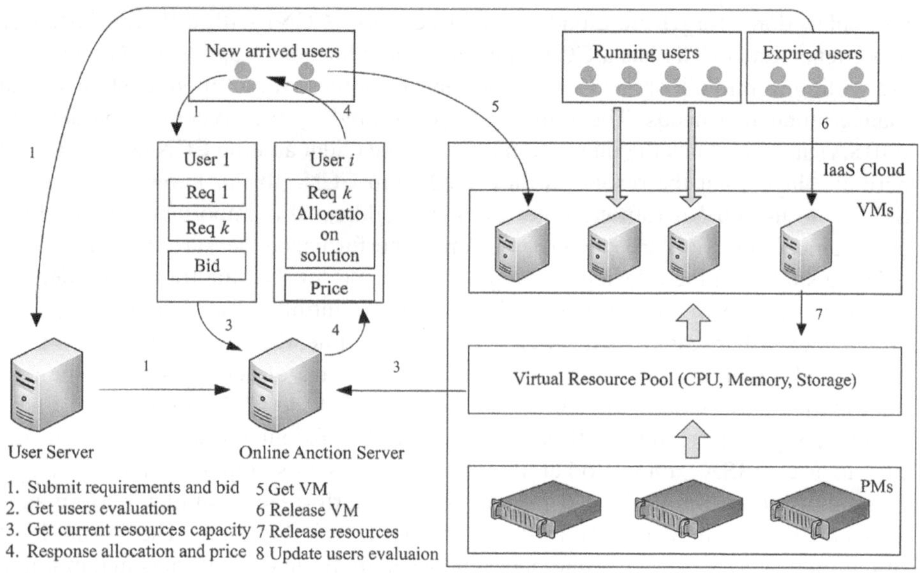

FIGURE 10.1 System architecture.

with diverse needs can join CCN on demand and need different types of servers at the same time. Our model, inspired by the Proxy Reseller (PA) source of trust between consumers and relevant CCNs, covers this. Each client submits a request to disable PAs; Finally, PAs try to meet demand with products available from CCN. Each CCN operates under the control of a CCN leader who interacts with the CP and PA. Due to different plan types and personal values, customers will have different options for different servers (hybrid), assuming this is disclosed to their PA. In this model, PA and CCN managers use competing language to express the customer's needs and assessments. However, discussion of competing languages is beyond the scope of this chapter. Interested readers can get more information from there. In addition, there is recent research on the creation of privacy-preserving auctions for spectrum allocation in networks, and this research attempts to use cryptographic techniques for combinatorial auctions and double auctions to protect domain information and bidders' prices. In these drawings, we want the customer's price matching and information to be shared directly with the relevant PA. Masking these statistics requires the use of good encryption techniques to suit our campaign plans left for the future. In this Figure, we show the basis of CCN executives renting multiple servers from CPs through joint ventures in a single market: trade show stores and the entire business as a private sale. In a flat rate market, CPs offer their servers at a hard and fast price. In commercial sales where CCN administrators are bidding competitively for the simultaneous acquisition of wireless servers that support CP pricing, CPs offer their servers with provisioning costs (for example, the lowest estimate of the relevant customer lease). Market parity is suitable for server rentals with long downtime. In this case, CP intends to rent server at a fixed price so that you don't have to use the server in the future However, the market managed by open sales is better for servers with less downtime. In this case, CP also wants to provide its own servers in the future. Wi-Fi, CP provides all services by lowering server price so you can surf faster In addition, CCN executives who need high-value Wi-Fi wireless and want to rent it for a long time join the fixed price market, while CCN executives reduce participation in sales-based marketing as a whole. As can be seen from the plan, there are stages involved in collecting and selling resources. While key documents capture interactions between PAs and CCN executives, they represent most of the CCN's leads and CP interactions while they represent CU's represents competitive pricing and nonpublic proprietary information below we discuss and examine these lines.

10.8 SYSTEM ANALYSIS

Figure 10.2 depicts the detailed workflow of the anomaly detection module, highlighting the three main stages: data preprocessing, feature extraction, and classification. This workflow ensures accurate detection of irregular patterns in real-time data streams. We consider the market, where the bids' range is adopted from the history of winners' bid of memory optimized instance (x1.32xlarge) in the past three months. The minimum (maximum) value of bids corresponds to the 50% of the minimum winner bid (maximum winner bid) in that dataset upon requesting the cloud instance for 24 hours. The presented analytic results for the CCN manager's income and the results obtained from 10000 Monte Carlo simulations are depicted It can be seen that the analytic results and the simulation results agree well with each other. Also it can be observed that the CCN manager's income becomes less sensitive to the number of PAs after the market gets sufficiently crowded. Moreover, the effect of increasing the patience time on the CCN manager's income becomes more pronounced when

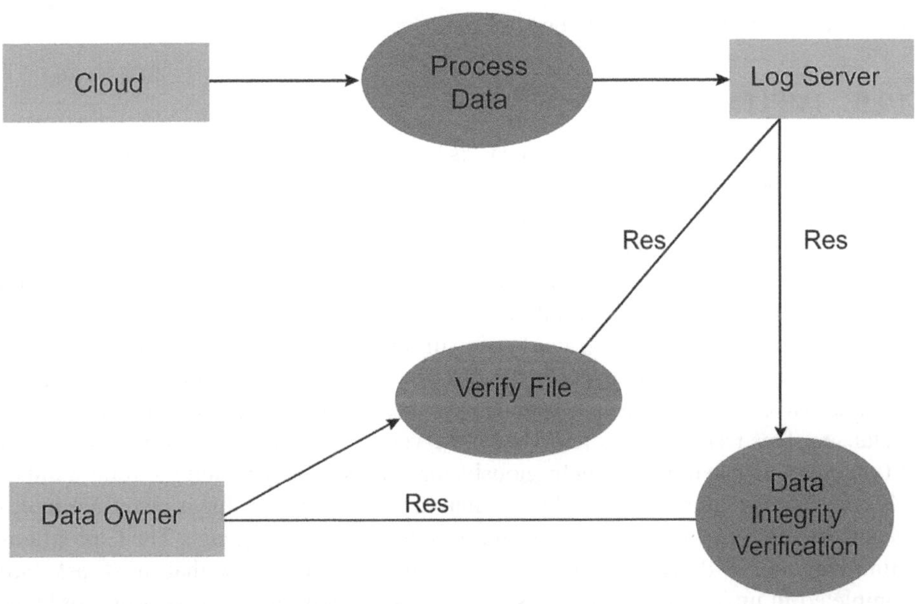

FIGURE 10.2 System work flow.

there are fewer participant PA's. The improvement that can be obtained using the OBSAs instead of the sequential combinatorial auctions is shown in terms of the total number of winners and total sold cloud servers. For this simulation, ten types of servers are considered where the CCN manager holds 1,000 auctions for each of these types synchronously at every minute, where the total number of needed types of servers for each PA is uniformly distributed in the interval. The arrival rates (per minute) for servers are uniformly distributed in the interval [1/15, 1]. This interval is derived based on the reported data in, which indicates that the majority of task executions are under 15 minutes. It is assumed that customers do not leave the market without obtaining their demanded resources (long patience time). The results indicate an increase in the customers' satisfaction, utilization of resources, and the CCN manager's income upon using the OBSAs. MODULES In this Proposed System, there are four modules. They are:

- Data owner
- Log Server
- Cloud Server
- End User

Data Owner: In this module, the data owner uploads their data in the cloud server. For the security purpose the data owner encrypts the file.

End User: In this module, user logs in by using his/her username and password.

Cloud Server: The cloud server manages a cloud to provide data storage service. Data owners encrypt their data files and store them in the cloud for sharing with remote user.

Log Server: In this module, Log Server log in by using his/her username and password. After Login he will do the operations.

10.9 IMPLEMENTATION

We remember the market and the forecast is based on the winning data of the improved memory in the last three months. The minimum (maximum) price of the bid then corresponds to 50% of the minimum bid (maximum bid), for example 24 hours after the cloud wind request. The analysis results of the CCN leader returns provided by and the results from 10,000 Monte Carlo simulations are plotted and it can be seen that the analysis results and simulation results are displayed in all variables. It has also been found that once the market is sufficiently crowded, the income of CCN executives becomes less relevant to the PA. The improvements available from OBSA compared to connected products spanned across all earnings types and averages of the instances. This test considers ten types of servers where the CCN administrator runs 1,000 auctions per minute synchronously for each server type and the total number of server types required by each PA is equal in C. The server value is also (in minute) id and is evenly divided between the duration of the script language. The C programming language is derived from the above statistics which show that most tasks are completed in under 15 minutes. It is generally accepted that customers do not leave the business without getting the resources they need (battery life). The results showed that customer satisfaction, asset utilization, and CCN executives' results improved after implementing OBSA. The main task of the client is to move the login window to the personal window. This module is designed for support saving. In this login page, we need to enter the login ID and password. It will check if username and password match (male or female id and password are valid). If we enter an incorrect username or password, we will not be able to access the login window in the user window, this will cause an error message. Therefore, we prevent unauthorized users from accessing the login window. It will provide great security for our business. So the server has an identity role and a server password as well as checking the authenticity of the role. It effectively improves security and prevents unauthorized users from accessing the network. Here we check the login and server authentication. The owner will test its content and enter the web page where he will add the details of the product he wants to sell, upload the product information and the average price of the product. Similarly, many business owners will have access to their information. Customers will log in and log in with an ID and password. The customer will look at the items uploaded using the owner and claim the items by price and location or location and can be a price negotiation after receiving customer feedback.

10.10 CONCLUSION

In this work, we proposed a comprehensive two-stage framework to describe resource allocation and gathering in modern cloud networks. The first stage describes the interactions between the PAs and the CCN managers. For this stage, OBSAs along with their theoretical analysis are proposed, which enjoy a simple winner determination process and provide the truthfulness property. The second stage models the interactions between the CCN managers and the CPs. For this stage, a theoretical framework is developed to model the bidding behavior of the CCN managers.

10.11 FUTURE ENHANCEMENT

For future work, one worthwhile direction is to explore the optimization of the social welfare or other parameters of interest. Studying the resource allocation and the load-balancing problems jointly is also interesting. In this case, a CCN manager should consider the geographical locations of the servers and CPs to find the optimal resource allocation.

REFERENCES

[1] S. Wu, M. Liwang, D. Wang, X. Wang, C. Wu, J. Tang, and Z. Jiao, Effective two-stage double auction for dynamic resource trading in edge networks via overbooking, *arXiv preprint arXiv:2501.04507*, 2025.

[2] Microsoft. (2017) Microst azure. [Online]. Available: https://azure.microsoft.com/en-us

[3] Y. Zhang, D. Niyato, P. Wang, and E. Hossain, Auction-based resource allocation in cognitive radio systems, *IEEE Commun. Mag.*, vol. 50, no. 11, pp. 108–120, Nov 2012.

[4] A. W. Min, X. Zhang, J. Choi, and K. G. Shin, Exploiting spectrum heterogeneity in dynamic spectrum market, *IEEE Trans. Mobile Comput.*, vol. 11, no. 12, pp. 2020–2032, Dec 2012.

[5] L. Gao, Y. Xu, and X. Wang, MAP: Multi- auctioneer progressive auction for dynamic spectrum access, *IEEE Trans. Mobile Comput.*, vol. 10, no. 8, pp. 1144–1161, Aug 2011.

[6] L. Gao, X. Wang, Y. Xu, and Q. Zhang, Spectrum trading in cognitive radio networks: A contract-theoretic modeling approach, *IEEE J. Sel. Areas Commun.*, vol. 29, no. 4, pp. 843–855, April 2011.

[7] X. Wang, Z. Li, P. Xu, Y. Xu, X. Gao, and. H. Chen, Spectrum sharing in cognitive radio networks An auction-based approach, *IEEE Trans. Syst., Man, Cybern., Part B (Cybernetics)*, vol. 40, no. 3, pp. 587–596, June 2010.

[8] W. Wang, B. Liang, and B. Li, Revenue maximization with dynamic auctions in IaaS cloud markets, in *Proc. IEEE/ACM 21st Int. Symp. Quality Serv. (IWQoS)*, June 2013, pp. 1–6.

[9] S. Jha, N. Jha, D. Prashar, S. Ahmad, B. Alouffi, and A. Alharbi, Integrated IoT-based secure and efficient key management framework using hashgraphs for autonomous vehicles to ensure road safety, *Sensors*, vol. 22, no. 7, p. 2529, 2022.

[10] Google. (2017) Google cloud platform. [Online]. Available:https://cloud.google.com/

[11] J. Zhu, Z. Zhang, B. Wang, K. Shao, Z. Hetu, and X. Wang, FDTA: A value-guaranteed fair data trading approach with a three-stage Stackelberg game, *IEEE Internet of Things J.*, 12(8), 9965–9977. 2025.

[12] D. Bernstein, E. Ludvigson, K. Sankar, S. Diamond, and M. Morrow, Blueprint for the inter cloud – protocols and formats for cloud computing interoperability, in *Proc. 4th Int. Conf. Internet Web Appl. Serv.*, May 2009, pp. 328–336.

[13] S. Jha, S. Routray, H. A. M. Abdeljaber, and S. Ahmad, A novel approach for decision support system in cricket using machine learning, *Int. J. Comput. Appl. Technol.*, vol. 69, pp. 86–92, Oct. 6, 2022.

[14] Multiple resource procurement in cloud computing, in *Proc. 12th Innov Conf. Intell. Syst. Design Appl.*, 2012, pp. 337–344.

[15] U. Lampe, M. Siebenhaar, A. Papageorgiou, D. Schuller, and R. Steinmetz, — Maximizing cloud provider profit from equilibrium price auctions, in *Proc. IEEE 5th Int. Conf. Cloud Comput.*, June 2012, p. 83.

[16] Amazon. (2017) Amazon ec2 spot instances. [Online]. Available: https://aws.amazon.com/ec2/spot/

[17] S. Jha, S. R. Shakya, S. Ahmad, and Z. Zhou, editors, *Quality Assessment and Security in Industrial Internet of Things*. CRC Press, Taylor & Francis Group, 2025.

[18] S. Jha, Model Selection Procedure in Alleviating Drawbacks of The Electronic Whiteboard, in *2021 3rd International Conference on Advances in Computing, Communication Control and Networking (ICAC3N)*, Greater Noida, India, 2021, pp. 319–322, doi: 10.1109/ICAC3N53548.2021.9725365.

[19] S. Jha, A. Sultan, M. Alharbi, B. Alouffi, and S. Sebastian, Secured and provisioned access authentication using subscribed user identity in federated clouds, *Int. J. Adv. Comput. Sci. Appl.*, vol. 12, no. 11, 2021.

[20] S. Ahmad, and M. M. Afzal, Deployment of fog and edge computing in IoT for cyber-physical infrastructures in the 5G era, in *Sustainable Communication Networks and Application: ICSCN 2019,* Springer International Publishing, 2020, pp. 351–359.

11 Smart Solutions for a Smarter Nation

Cloud Computing's Role in Nepal's Development

Sudan Jha

11.1 CONTEXT OF TECHNOLOGICALLY DEVELOPING COUNTRIES

Technologically developing countries like Nepal are at a pivotal moment in their digital transformation journey. With challenges such as limited infrastructure, access to technology, and a fragmented educational landscape, the path to development often appears obstructed. However, cloud computing offers a unique opportunity to overcome many of these hurdles by providing scalable, cost-effective technological solutions that do not require large upfront capital investments in infrastructure.

In Nepal, a country known for its stunning natural beauty, there is a pressing need to embrace technological advancements that can bridge the developmental gap in sectors such as education, healthcare, agriculture, and governance. Cloud computing, through its ability to deliver services over the internet, enables access to cutting-edge technologies without the need for expensive physical hardware or a high level of initial investment. It allows the people of Nepal, from urban centers to remote villages, to access the digital world more effectively and affordably.

Cloud computing has the potential to transform industries and sectors by providing them with scalable, on-demand computing resources. For a country like Nepal, where physical infrastructure is often lacking, cloud computing can act as an enabler, bringing solutions that are both accessible and sustainable in the long run.

Nepal's digital divide is stark: only 38% of rural households have internet access compared to 72% in urban areas (Nepal Telecommunications Authority, 2023). This disparity exacerbates challenges in education, healthcare, and economic mobility. For instance, in Humla District, students trek hours to reach schools with intermittent electricity, rendering cloud-based tools unusable without offline alternatives. However, initiatives like the Government of Nepal's Digital Nepal Framework aim to bridge this gap by 2030 through partnerships with ISPs and cloud providers like Huawei Cloud. These collaborations prioritize last-mile connectivity using low-orbit satellites and community Wi-Fi hubs, as piloted in Kavre District in 2022. Such efforts underscore cloud computing's role not just as a tool, but as a catalyst for equitable development.

DOI: 10.1201/9781003609773-11

11.3 THE POTENTIAL OF CLOUD COMPUTING IN NEPAL

For Nepal, cloud computing presents significant opportunities across several key sectors:

11.3.1 Cloud Computing in Education

Education in Nepal faces significant challenges, particularly in remote areas where access to quality education and digital resources are limited. Cloud computing can bridge these gaps by offering scalable solutions such as Learning Management Systems (LMS), virtual classrooms, and collaborative tools that do not require heavy physical infrastructure.

Example: Cloud-based platforms like Google Classroom and Microsoft Teams have been used to provide online education across schools in Nepal. These platforms allow for seamless collaboration, access to educational materials, and even student assessment in a digital environment, making education more accessible in remote areas.

Despite cloud education tools' promise, Nepal faces implementation hurdles. In 2022, a UNICEF survey revealed that 68% of rural teachers lacked training to use platforms like Google Classroom. Power outages averaging 8 hours daily in districts like Mugu further disrupt access. Solutions are emerging:

- **Offline Hybrid Models:** *OLE Nepal*'s Kolibri app syncs with the cloud when online, caching lessons for offline use.
- **Community Centers:** Villages in Dolakha District share solar-powered tablets with preloaded Khan Academy videos.
- **Teacher Training:** The *Digital Education Action Plan* (2023) budgets $2 million to certify 5,000 educators in cloud tools by 2025 [5].

The key educational advantages enabled by cloud computing—particularly in terms of cost efficiency, accessibility, scalability, and collaborative learning—are summarized in Table 11.3.

11.3.2 Cloud Computing in Healthcare

Cloud computing plays a vital role in modernizing healthcare systems by offering cloud-based Electronic Health Records (EHR) systems, telemedicine platforms, and

TABLE 11.3
Benefits of Cloud Computing in Education

Benefit	Description
Cost Efficiency	Reduced infrastructure and maintenance costs.
Accessibility	Access to educational materials from any location.
Scalability	Ability to scale the infrastructure to accommodate more students.
Collaboration	Facilitates collaborative learning and communication.

data storage solutions. This allows healthcare providers to deliver better services while reducing operational costs.

Example: Nepal can adopt cloud-based healthcare systems for remote diagnosis and patient management, which is crucial for the rural population who may not have immediate access to specialized healthcare.

Cloud-based healthcare in Nepal hinges on data security and connectivity. Patan Hospital's adoption of CloudCare EHR reduced patient wait times by 40%, but concerns persist. Nepal lacks HIPAA-like laws [6], leaving sensitive data vulnerable. In 2023, the Health Tech Policy Draft proposed encryption standards and penalties for breaches. Meanwhile, telemedicine pilots in Mustang use lightweight apps like Telehealth Nepal to compress data for low-bandwidth areas, proving innovation thrives under constraints [7].

11.3.3 Cloud Computing in Agriculture

Nepal's agricultural sector, which forms the backbone of its economy, can greatly benefit from cloud computing. Using cloud-based agricultural management systems, farmers can monitor crop health, receive weather updates, and optimize resource use. The integration of Internet of Things (IoT) devices in agriculture, paired with cloud computing, can enhance precision farming practices.

Example: Cloud-based platforms can provide farmers in Nepal with real-time weather forecasts and crop performance data, empowering them to make informed decisions about irrigation and pesticide use [8].

As illustrated in Table 11.4, cloud-based applications such as weather forecasting, crop monitoring, and supply chain management play a critical role in modernizing Nepal's agricultural practices.

11.4 SUSTAINABILITY AND CLOUD COMPUTING

Sustainability in cloud computing refers not only to the environmental benefits, but also to the long-term financial and operational efficiency that cloud services provide to developing countries like Nepal.

11.4.1 Environmental Sustainability

Cloud providers are increasingly focusing on reducing their environmental footprint by using renewable energy for their data centers. For Nepal, a country with abundant

TABLE 11.4
Cloud Computing Applications in Agriculture

Application	Description
Weather Forecasting	Provides accurate weather predictions for farmers.
Crop Monitoring	Uses cloud-based sensors to monitor soil conditions and crop health.
Supply Chain Management	Streamlines the supply chain from farms to markets.

natural resources, adopting green cloud solutions can reduce the nation's carbon footprint.

Example: Cloud computing providers like Amazon Web Services (AWS) and Google Cloud have committed to carbon-neutral operations. Nepal could collaborate with these providers [9] to reduce its own emissions.

11.4.2 Economic Sustainability

The cost-efficiency of cloud computing offers financial sustainability by reducing the need for significant capital investments in physical infrastructure. This is particularly important for a developing economy like Nepal, where financial resources are limited.

To highlight the economic sustainability of cloud adoption, Table 11.5 compares upfront costs, maintenance expenses, and energy consumption between traditional IT infrastructure and cloud computing models. Figure 11.1 depicts the bar graph comparing the costs for Traditional IT versus Cloud in the categories of Upfront Capital Costs, Annual Maintenance, and Energy Consumption. The graph clearly illustrates the significant cost difference, with cloud computing being much more cost-efficient across all categories.

11.5 CHALLENGES IN CLOUD ADOPTION IN NEPAL

While the potential benefits of cloud computing in Nepal are vast, there are several challenges that must be addressed for successful adoption.

11.5.1 Infrastructure and Connectivity

Despite progress, Nepal still struggles with issues related to internet bandwidth, network reliability, and the availability of high-speed internet in rural regions [10]. This is a significant barrier to cloud adoption, as reliable internet connectivity is essential for cloud computing.

TABLE 11.5
Cost Comparison: Traditional IT Infrastructure versus Cloud Computing

Expense Category	Traditional Infrastructure	Cloud Computing
Upfront Capital Costs	High (servers, storage, etc.)	Low (subscription-based)
Maintenance Costs	Ongoing (staff, hardware upgrades)	Minimal (outsourced to provider)
Energy Consumption	High (on-site data centers)	Low (shared resources)

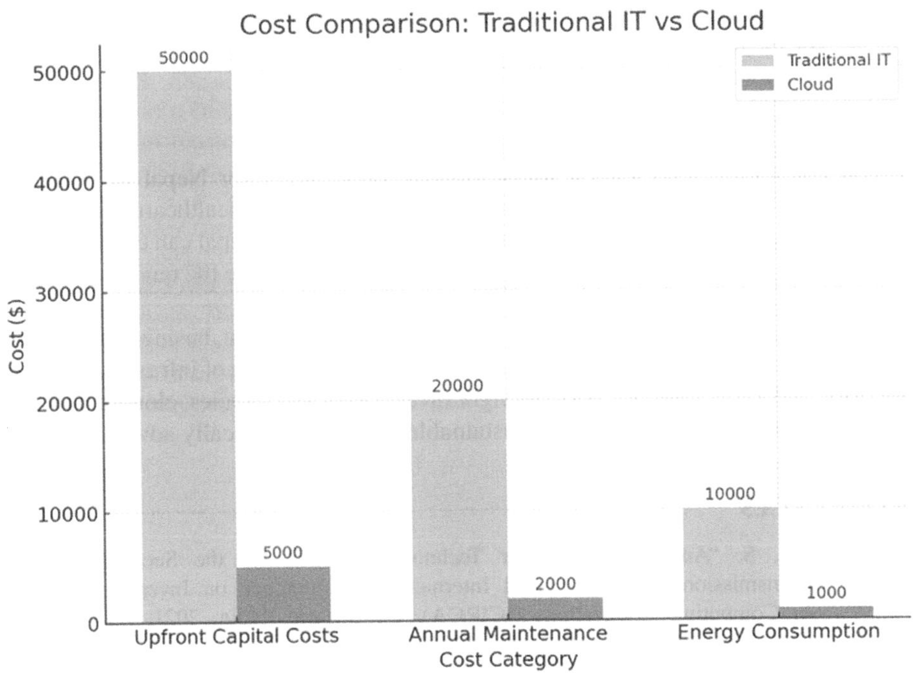

FIGURE 11.1 Cost comparision: Traditional vesus cloud.

11.5.2 Data Security and Privacy

With the adoption of cloud computing, there is a rising concern over data privacy and security [11]. Nepal needs to develop appropriate policies and infrastructure to ensure the protection of sensitive data, especially in sectors like healthcare and finance [12].

11.5.3 Skill Development

There is a growing demand for cloud computing professionals in Nepal, but the country faces a shortage of skilled workers [13] in this field. Investments in education and training programs will be essential for building local capacity and expertise.

Nepal's cloud adoption requires policy and collaboration shifts. The *Data Sovereignty Bill* (2024 draft) mandates that citizen data stored offshore (e.g., AWS Singapore) must mirror locally, raising costs. Conversely, India's *MeghRaj* cloud offers a regional alternative with Nepali-language support. Skills gaps are being addressed:

- **Academic Partnerships:** Tribhuvan University's *BSc in Cloud Engineering* (launched 2023) has 200 enrollees.
- **Private Sector:** CloudSewa's *#SkillUpNepal* program trains 500 youths annually in AWS/Azure.

12 Exploring the Role of Digital Twins in Urban Planning

Opportunities and Challenges for Smart Cities

K. Ravi Kishore and Taarani Kaligotla

12.1 INTRODUCTION

Urbanisation is happening at a pace never seen before, and it poses new challenges and opportunities for city administrators and planners. As populations expand, so do the complexities of handling urban infrastructure, maintaining public safety, and sustaining urban environments. Innovative technologies are therefore imperative in solving these complex problems. One technology that emerges in this zone is the use of digital twins.

12.1.1 Overview of Digital Twins

Digital twins are virtual representations of physical objects that reflect the state and behaviour of their real-life counterparts through real-time data. This term was first mentioned in the manufacturing industry, where digital twins were deployed to simulate machinery and production lines. Through the continuous generation of data from sensors and Internet of Things (IoT) devices, digital twins parallelly change with their counterparts, which are present in the real world; this enables active monitoring and assessment of data. The main reason for having a digital twin is its ability to mirror its physical counterpart accurately, reflecting any changes made actively and predicting future states on the basis of the current data.

In manufacturing, digital twins have changed production lines by facilitating predictive maintenance. This tactic reduces downtime and increases operating efficiency by anticipating equipment failure before it takes place so that appropriate interventions can be made. The medical industry has also considered digital twins to be an efficient way of simulating a patient's body that needs to be assessed for treatment, which helps in personalised treatment plans and better patient outcomes. For example, a digital twin of a patient's heart can model the effect of various treatments and help doctors understand the most suitable method. These applications prove the general-purpose nature of digital twins and make their use in urban planning and smart city development especially interesting.

DOI: 10.1201/9781003609773-12

FIGURE 12.1 Digital twins technology for smart city development.

As cities grow to become more complex, they tend to require more sophisticated tools to monitor and enhance urban functioning. Digital twins successfully accomplish this by presenting a complete virtual model of the city's infrastructure, systems, and environment. This virtual model is updated in real time using information from disparate sources such as traffic cameras, environmental sensors, and utility management systems. By combining and processing this data, digital twins can offer meaningful insights and facilitate data-driven decision-making in urban planning and management.

Figure 12.1 illustrates the conceptual representation of a digital twin, showing how real-time data from physical urban systems is integrated into a virtual city model for monitoring, simulation, and decision-making.

12.1.2 Importance in Smart City Development

A smart city is an urban area that utilises digital technology and data collection to enhance the quality of life for its residents, improve city operations, and promote sustainability. Digital twins are the key to making an urban city possible. They allow city planners and administrators to observe, control, and optimise various aspects of the city, ranging from infrastructure and traffic to energy usage and public safety.

One of the most important advantages of digital twins in developing smart cities is that they can model various situations and forecast the results. For example, by simulating a scenario like the flow of traffic in the city, digital twins can help the city planners in understanding bottleneck areas and taking measures to lessen traffic. By doing so, commute times are lowered and emissions are reduced, making the city greener and more efficient. When considering an infrastructure perspective, digital twins give infrastructure a new way to deal with things. With continuous monitoring of the condition of infrastructure like bridges, roads, and buildings, administrators can predict failures and schedule maintenance work before issues even arise. This proactive strategy helps the infrastructure industry in many ways. It also helps to lower maintenance costs, minimising turmoil in city life.

Public safety is another key sector where digital twins have the potential to contribute significantly. With data assimilation of surveillance cameras, sensors, and emergency response systems, digital twins offer end-to-end visibility into the city's security situation. This helps in rapid identification and response to security threats, making residents safer. Digital twins can also be used to assist predictive policing by detecting likely hotspots for crime and enabling law enforcement to allocate resources better.

Digital twins are also important in ensuring environmental sustainability. Through improving waste management systems, energy usage, and resource distribution, they help cities to reduce the environmental impact caused by them.

By incorporating digital twins into the development of smart cities, we receive many benefits, such as better resource management, increased public safety, and increased sustainability. By giving the city planner a virtual replica of the city to be developed, the planners can make efficient choices that improve the quality of life for citizens and maximise the city's efficient operation.

12.1.3 Objectives of the Chapter

This chapter aims to understand and show the potential of digital twins with respect to smart city development. It discusses the technological innovations behind digital twin technology, including data harvesting mechanisms, sophisticated analytics, and cloud computing. Through a deep dive into these tools, we can successfully discover how digital twins produce precise and dependable virtual replicas of urban cities.

The most critical part for the proper functioning of a digital twin is the mechanism that it uses to collect data. Urban infrastructure is equipped with sensors and IoT devices that collect real-time information on several parameters, including traffic flow, energy usage, and environmental status. This information is then combined and analysed to produce a digital model that keeps changing over time with the physical one. Machine learning (ML) and predictive analytics are employed to analyse the data and show proactive intelligence. Cloud computing offers the infrastructure necessary to store and process the huge datasets needed by digital twins in an efficient and cost-effective manner with scalability and flexibility.

In addition to this, the chapter will also focus on how different advantages of digital twins are brought to bear to improve smart city infrastructure. These advantages include optimised traffic management, enhanced security and surveillance, effective waste management, and efficient urban energy management. All these aspects are essentially giving rise to a more sustainable and productive urban landscape.

Yet, the deployment of digital twins is not without issues. The chapter will also cover the technical, operational, financial, and regulatory challenges to urban planners when embracing this technology. Technical challenges are the issues that deal with data accuracy, integration, and technical constraints. Operational challenges deal with the integration of digital twins into existing urban systems. Monetary challenges include the investment and expenses involved in digital twins. Regulatory challenges deal with issues like data privacy and security.

By recognising these challenges, we can suggest the measures to defeat them and ensure the successful implementation of digital twins in the context of smart cities. These measures involve the use of advanced data analytics solutions, middleware, cloud computing, and complete cost-benefit analysis.

To conclude, the chapter will reflect upon the future prospects and scope of digital twins. New trends, like improvements in artificial intelligence (AI) and 5G network integration, will continue to advance digital twins' abilities. AI and ML will improve the predictive and analytical abilities of digital twins, while 5G will support quick transfer of data and better real-time updates. The chapter ends with a vision of the future, highlighting the importance of ongoing research and development to unlock the full transformative potential of digital twins in urban planning and smart city building.

12.2 FUTURE PROSPECTS OF DIGITAL TWINS IN SMART CITIES

12.2.1 Enhancing Urban Infrastructure

Urban infrastructure may be rejuvenated by digital twins through the delivery of high-grained, real-time information about the condition and performance of city infrastructure assets. These include buildings and bridges, roads, and public utilities.

12.2.1.1 Real-Time Monitoring

Real-time monitoring using digital twins is the continuous collection and analysis of information from sensors incorporated within different components of infrastructure. The sensors offer constant real-time streams of data about a number of parameters, such as the stress level, temperature, vibrations, and status of the structure. City managers have the opportunity to gain real-time visibility of the status and performance of key infrastructure elements like roads, bridges, buildings, and public buildings. This continuous flow of information is crucial in the detection of potential faults before they can develop into major problems, hence enhancing the overall effectiveness and safety of urban infrastructure.

12.2.1.2 Predictive Maintenance

Predictive maintenance using digital twins redefines the landscape of urban infrastructure management, as it enables city planners to anticipate potential failure or degeneration of infrastructure components. With predictive analytics, digital twins analyse data collected from sensors embedded in various aspects of infrastructure, such as roads, bridges, buildings, and public buildings. The data-driven approach to monitoring provides real-time information regarding the condition and performance of the assets in real time, supporting proactive scheduling of maintenance. The result is the removal of significant downtime and an extension of infrastructure lifetime, finally resulting in improved efficiency and safety in cities.

12.2.1.3 Optimisation of Resources

The concept of the digital twin is to create a virtual replica of physical infrastructure that continuously collects and analyses information with sensors placed within the infrastructure. This technology presents a paradigm-shifting method of resource management to enable the maximisation of use of materials, manpower, and budgets within urban settings. With precise determination of the areas of priority and those that can wait until later, digital twins enable efficient management of resources, which has a pronounced effect of saving while enhancing the quality of service delivery. This mechanism helps cities sustain good infrastructure standards without going over budget or overallocating.

12.2.2 Energy Management and Sustainability

Digital twins are poised to transform the management of urban infrastructure through the provision of comprehensive, real-time information on the status and performance of city elements. Such components are roads, bridges, buildings, and public utilities.

12.2.2.1 Monitoring of Energy Consumption

Monitoring energy consumption is a significant application of digital twin technology for smart city development. Digital twins are able to monitor and analyse patterns of energy consumption in various sectors of the city at all times, which enables a clear understanding of how energy is being consumed in various industries. Such comprehensive monitoring enables inefficiencies to be identified and the areas where energy can be utilised to the maximum. This involves shifting non-peak energy consumption to off-peak times or installing demand-response programmes in order to ease the electrical grid at peak load periods. Using energy demand forecasting, cities are enabled by digital twins to allocate energy resources more efficiently, prevent grid overload, and reduce the use of expensive and emission-prone plants. Moreover, digital twins may help integrate renewable energy sources by forecasting variability in energy supply using solar or wind power and adapting energy usage accordingly.

12.2.2.2 Renewable Energy Integration

Renewable energy integration is an essential aspect of urban sustainable energy management, and the role of digital twins is highly important in this regard. Digital twins enable the integration of renewable energy resources such as solar and wind into the city's energy grid by simulating both energy production and consumption in real time. These models allow cities to manage supply and demand effectively and provide a stable and efficient energy supply. With the analysis of weather patterns and energy production from renewable sources, digital twins can predict variations in energy availability. In short, digital twins facilitate the smooth integration of renewable energy into urban grids and create a sustainable, resilient, and green urban energy system.

12.2.2.3 Smart Grids

Smart grids are a revolutionised method of managing urban energy, and digital twins play a critical role in their development and operation. They have the capacity to adapt dynamically to varying energy demand and supply changes, providing efficient and reliable energy distribution. Digital twins' ability to adjust dynamically in real time according to consumption patterns is one of the most prominent benefits of using digital twins in smart grids. Through ongoing monitoring of energy consumption throughout the grid, digital twins can fine-tune electricity distribution, lower peak loads, and minimise energy loss which prevents the overloading of the grid and enhances overall system efficiency. For instance, smart grids can promote higher energy consumption or storage in times of high renewable energy output and lower demand when production is low.

12.2.3 Transportation and Mobility Solutions

Mobility and transport are critical components of urban living, and digital twins introduce innovative solutions to enhance these systems.

12.2.3.1 Traffic Management

Digital twins can simulate traffic flows and predict where congestion will happen, and therefore effective traffic management strategies can be put in place. This involves controlling traffic light cycles, providing alternative routes, and optimally controlling road closures. For example, during large events, digital twins can model the effect of traffic and recommend ways to achieve the least disruption. They assist in planning road networks that have the least congestion and optimise overall traffic flow. Digital twins can also forecast traffic trends in the future based on real-time data, enabling proactive adjustments and long-term urban planning. By providing real-time intelligence and predictive analytics, digital twins help create wiser and more responsive transport systems, improving mobility and reducing journey times for city residents.

12.2.3.2 Public Transport Optimisation

Digital twins enable public transport systems to be optimised with the help of data analysis. This involves optimising bus and rail schedules based on real-time demand, route optimisation, and making the entire public transport system more efficient. Digital twins can also be used to schedule new transit routes and expand existing networks. For example, if a particular bus passes through a common phenomenon of overcrowding, there can be suggestions by digital twins to add frequency to buses or offer additional routes to ease the load.

12.2.3.3 Autonomous Vehicle

Digital twins can prove to be very valuable in designing and deploying autonomous vehicles. Through the digital twin, which imitates the real world, autonomous vehicle technology is able to be tested and certified in various conditions, thus becoming safer and more dependable. This is done by simulating various levels of traffic flow, weather, and possible roadblocks in order to ensure autonomous vehicles can reach their destinations safely and efficiently.

12.2.4 Emergency Response and Public Safety

Digital twins support public safety and emergency response operations in smart cities with substantial assistance.

12.2.4.1 Monitoring and Surveillance

By fusing data from surveillance cameras, sensors, and other monitors, digital twins create a unified view of public safety. This enables quicker detection and action for events such as accidents, criminal activities, and natural disasters. For example, digital twins can analyse streams from multiple cameras to detect unusual activities or monitor crowd movement in cases of mass gatherings to ensure public safety.

12.2.4.2 Emergency Response Planning

It is possible to use digital twins to model different emergency scenarios, including fires, floods, and terrorism, for enhancing response planning. Simulation of the probable effects of such disasters allows city planners to create better evacuation plans, dispatch emergency resources, and reduce casualties and damage. For instance, digital twins can assist in identifying the best evacuation routes and emergency services staging locations for a catastrophic event.

12.2.4.3 Risk Assessment

Digital twins allow constant risk assessment by analysing data from various sources. They comprise identifying areas prone to natural disasters, identifying the vulnerability of the critical infrastructure, and monitoring public health risks. For example, digital twins can make predictions for flood risks based on weather patterns as well as drainage systems' capacity so that precautionary steps can be taken to protect vulnerable areas. They can also help analyse the impacts of climate change on urban infrastructure and planning and long-term resilience.

12.2.5 Health and Human Services

With the integration of digital twins within human and health services, the provision and efficiency of such critical services can be greatly improved.

12.2.5.1 Health Facilities Management

Digital twins can duplicate the operation of health facilities in order to utilise resources such as medical equipment, staff, and space optimally. This reduces the time taken for waiting, enhances patient care, and increases the overall efficiency in operations. For instance, digital twins might replicate patient flow through a hospital to determine bottlenecks and recommend changes in personnel or facility design.

12.2.5.2 Public Health Monitoring

Through the analysis of data from health institutions such as hospitals, clinics, and public health departments, digital twins can provide useful information about the trends in public health. This includes monitoring the incidence of infectious diseases, monitoring vaccination, and areas where there are high rates of chronic conditions. For instance, digital twins can assist in forecasting and controlling outbreaks of diseases like influenza through analysing data on patient symptoms, patterns of mobility, and environmental conditions.

12.2.5.3 Social Services

Digital twins can assist in the provision of social services through the simulation of the needs and demographics of various communities. This supports planning and allocating resources accordingly to ensure that vulnerable groups get the necessary assistance. For instance, digital twins can establish areas of high food insecurity and inform the allocation of relief. They may also assist in streamlining the delivery of services like housing programmes, mental health services, and employment training programmes.

12.3 CHALLENGES IN URBAN PLANNING WITH DIGITAL TWINS

12.3.1 Data Management and Integration

In order to create digital twins for urban city planning, one of the most important things to keep in mind is being able to manage and integrate data. This can be quite challenging because of the size, diversity, and complexity of the data.

12.3.1.1 Volume and Variety of Data

To build a detailed virtual replica of a city, digital twins rely on data from multiple sources. This includes sensors in infrastructure, IoT devices, historical datasets, and live data from monitoring systems. The data covers many areas, such as traffic patterns, energy use, environmental factors, and public safety metrics. The amount of data generated can be overwhelming; for instance, a single city can have thousands of traffic sensors that constantly send information. Large volumes of data need to be managed with advanced systems that efficiently process these vast datasets. Moreover, the wide range of data types and formats poses challenges since each system and sensor has its own rules on which it works. Integrating these various data streams into a single system results in good data processing and a robust infrastructure.

12.3.1.2 Data Quality and Precision

The success of a digital twin largely relies on the accuracy and quality of the data utilised. Reliable and accurate virtual models can be created with good-quality data. Faulty sensors, data entry errors, and inconsistencies in data sources can undermine accuracy. For digital twins to be reliable, rigorous data validation processes need to be implemented. This involves ongoing observation for accuracy and completeness, as well as mechanisms for the detection and prevention of errors or anomalies. Ensuring data is up to date is also important for delivering a true picture of reality. Resolving inconsistencies in data early helps to preserve the integrity of the digital twin, with predictions and simulations being made based on trustworthy data.

12.3.1.3 Data Silos

One of the biggest challenges in handling urban data is the existence of data silos. Often, important information needed for city planning is trapped in disconnected systems or handled by different departments, creating fragmented data that is hard to integrate. These silos can affect the flow of information between city departments and outside parties, lowering the efficiency of digital twins. Thus, it is necessary that these silos be removed in order to make the digital twin models work the best they can. This requires developing standard data formats and communication systems that improve collaboration across various systems and departments. With data management platforms, it becomes easier to collect and share data. Additionally, setting rules for combining data can enhance efficiency and cooperation among data sources, increasing access for everyone.

12.3.1.4 Data Management and Integration Approaches

A few methods that can manage and integrate data in digital twin models are:

- Leverage Smart Data Management Tools: Modern analytics and data management tools can help handle large volumes of data.
- Standardise Data Formats and Communication Protocols: Developing and applying standard data formats can result in better integration between different systems. Standardisation helps simplify data across various sources and ensures compatibility.
- Implement Strong Data Validation Procedures: Creating effective data validation and quality checks can make data in digital twins more accurate and reliable. Periodic checks and updates and data-cleansing automation software can assist in keeping data of top quality.
- Foster Data Integration and Sharing: Developing connected data management platforms and creating a data sharing culture among city departments and external organisations can cut silos. Collaborative efforts and shared data repositories make information more useful and accessible.
- Leverage Cloud Computing: Cloud computing provides the scalable storage and processing power needed to handle the large and diverse datasets used in digital twins. Cloud platforms also allow adjustments based on different data volumes and computing needs.

By addressing these challenges and implementing effective strategies, cities can improve how they deal with the vast amounts of data required for digital twins. This will result in the development of reliable digital twin models that improve urban planning and management.

12.3.2 Interoperability Issues

Interoperability issues are a significant concern when integrating digital twins with various systems and technologies. Ensuring seamless communication and functionality across different platforms requires addressing several key challenges.

12.3.2.1 Diverse Technologies

Digital twins leverage a range of technologies, each with its set of functionalities and requirements. These technologies typically include:

- **Internet of Things (IoT) Devices**: These sensors and devices collect real-time data from physical assets and environments, such as traffic sensors, weather stations, and energy meters.
- **Artificial Intelligence (AI) Algorithms**: AI and ML algorithms analyse the data collected to provide insights, make predictions, and simulate different scenarios.
- **Simulation Tools**: These tools create virtual models of physical entities, enabling the testing of various scenarios and strategies.

The challenge arises when trying to ensure that these diverse technologies work together seamlessly. For example, integrating an AI-based traffic management system with existing city infrastructure may require developing custom interfaces and middleware solutions. These custom solutions facilitate communication between different technologies, but they can be complex and costly to implement. Ensuring compatibility across various platforms and technologies requires a well-coordinated approach to system integration and communication.

12.3.2.2 Standardisation

A significant barrier to interoperability is the lack of universally accepted standards for digital twin implementation. This absence of standardisation affects several aspects:

- **Data Formats**: Different systems may use varying data formats, making it difficult to aggregate and analyse data from multiple sources.
- **Communication Protocols**: Without standardised communication protocols, systems may struggle to exchange information effectively, leading to integration challenges.
- **Security Measures**: Inconsistent security measures across platforms can create vulnerabilities and complicate the protection of sensitive data.

Developing and adopting industry wide standards is crucial for overcoming these interoperability challenges. Standardisation helps ensure that different systems and platforms can work together efficiently, reducing the complexity and cost of integration. Industry bodies, consortia, and standards organisations play a key role in establishing these standards and promoting their adoption across the sector.

12.3.2.3 Legacy Systems

Many urban environments are built on legacy systems—older technologies and infrastructure that predate modern digital twin solutions. These legacy systems often present several challenges:

- **Compatibility Issues**: Legacy systems may not be designed to interface with modern digital twin technologies, making integration difficult.
- **Resource Constraints**: Upgrading or integrating legacy systems can be resource-intensive, requiring significant time, effort, and financial investment.

- **Disruption Risks**: Implementing new technologies or making changes to legacy systems can risk disrupting essential services and operations.

Strategies for Enhancing Interoperability
To effectively address interoperability issues in digital twin implementation, several strategies can be employed:

1. **Develop Common Standards**: Industry-wide collaboration to establish and adopt common standards for data formats, communication protocols, and security measures can enhance interoperability. This may involve working with standards organisations and participating in industry consortia.
2. **Invest in Middleware Solutions**: Implementing middleware solutions can facilitate communication between diverse technologies and systems. Middleware acts as an intermediary, enabling different systems to interact and exchange information effectively.
3. **Promote Modular Design**: Designing digital twin systems with modularity in mind allows for easier integration with existing and future technologies. Modular components can be updated or replaced without overhauling the entire system.
4. **Engage in Collaborative Development**: Collaborating with technology vendors, standards organisations, and other stakeholders can help address interoperability challenges and drive the development of compatible solutions.
5. **Implement Interoperability Testing**: Regular testing of integrated systems and technologies can help identify and resolve interoperability issues early in the implementation process. This testing should include compatibility checks and performance evaluations.

By addressing the challenges of diverse technologies, standardisation, and legacy systems, cities can enhance the interoperability of digital twin solutions. This, in turn, will enable more effective integration, data sharing, and overall functionality, supporting the successful implementation of digital twins in urban planning and management.

12.3.3 Security & Privacy Concerns

The use of digital twins in urban planning raises serious security and privacy issues. This should be controlled carefully to enable the reliability and effectiveness of digital twin technologies while protecting sensitive data.

12.3.3.1 Data Security

Digital twins are based on large-scale data accumulation and processing, and thus data security is paramount. The information derived from sensor data of IoT, environmental sensors, and public safety systems is bound to be enormous and diversified in nature, and it is necessary to safeguard it against cyberattacks.

Major Points of Data Security Consideration:

- Encryption: Data encryption both at rest and in transit is crucial to avoid unauthorised access and ensure that sensitive data is kept confidential. Encryption algorithms are used to protect data by converting it into a format that can only be decoded by its owners with the help of decryption methods.
- Secure Communication Channels: Secure communication protocols (HTTPS or VPNs) make sure that the data is safe as it moves across various platforms like servers and sensors while data transfer takes place. Secure channels lower the possibility of interception and tampering.
- Regular Security Audits: Regular security audits help determine possible weaknesses in the digital twin models. The audits should cover penetration testing, code review, and risk analysis to fill security loopholes.
- Access Controls: Establishing strong access control measures provides the surety that only authorised members can access the data and manage it. Access controls include multi-factor authentication, role-based access controls, and audit trails that track and control access privileges. These sensors gather real-time data and send it to the central systems. It is necessary to have the data encrypted and protected with secure communication protocols so as to prevent unauthorised access, interception, and tampering.

12.3.3.2 Privacy Issues

The application of digital twins gathers and processes sensitive information regarding individuals, which in turn raises privacy concerns. This sensitive information may be the location, health, and personal preferences of an individual or an object; thus, this data must be processed carefully.
Key Considerations for Privacy Protection:

- Compliance with Privacy Laws: It is of utmost importance that the privacy laws are complied with, for example, the General Data Protection Regulation (GDPR) in the European Union or the California Consumer Privacy Act (CCPA) in the US.
- Data Anonymisation: Personal details can be masked by inculcating the methods of data anonymisation while the data transfer takes place. This method guarantees that even if data is obtained, it is not traceable to identifiable persons.
- Informed Consent: It is important to gain informed consent from people before gathering and processing their information. This means that the people get to choose whether they want in or out of the deal after they are told about how, where, and what information of theirs would be processed.

12.3.3.3 Risk of Data Breaches

The centralised nature of digital twin platforms, where large sets of data are processed, makes them prone to cyberattacks. Such an attack would result in the compromise of the nation's critical infrastructure, public safety, and private information.

An effective cyberattack can result in the manipulation of important information, for example, traffic management networks or energy grids, which basically means there are security risks. As such, it is important to develop and sustain secure digital twin setups to allow security threats to be evaded.

By taking care of data security, privacy, and data breach risk, cities can make digital twin technology more credible and trustworthy. By taking care of these processes, successful application of digital twins in city planning can be guaranteed, and a secure and robust digital infrastructure can be established.

12.3.4 Economic & Budgetary Constraints

The integration of digital twins in urban planning is a large-scale project that involves a lot of money and ongoing economic considerations. Proper handling of these constraints is a prerequisite for the successful implementation and feasibility of digital twin technologies.

The development and application of digital twins are linked with very high initial expenses. These can be divided into several key categories:

- Sensor installation: To develop an accurate digital twin, cities will have to install an IoT network of devices and sensors on various pieces of infrastructure. The sensors will give real-time data about traffic, weather, energy usage, etc. However, the cost of installing and managing these sensors is high.
- Infrastructure for Data Collection: The infrastructure needed to store and collect data is also expensive. This includes building data storage facilities, communication networks, and data processing centres. The infrastructure must be able to handle such large volumes of data that need to be stored and processed.
- Software Development: Creating the advanced software platforms that run digital twins is also an expensive task. It involves creating the digital models and receiving data from a wide range of sources.
- Testing and Integration: The digital twin models need to be integrated with the urban city systems, and then this integration must be tested for confirmation of successful working, which is a heavily costly task. It involves synchronising new systems with existing systems and overcoming potential integration problems.

It is not always easy to raise funds for these projects, especially in low-cost cities. Public agencies typically have to turn to grants, public-private partnerships, and innovative financing schemes in trying to fund these initial costs.

12.3.4.1 Ongoing Maintenance and Upgrades

After deployment, digital twins need constant updates, upgrades, and maintenance to work and be effective. These are ongoing costs:

- Hardware Maintenance: Sensors and other hardware need to be regularly maintained to make sure they are functional. Included in this are sensor calibration, replacement of faulty parts, and repair of physical wear and damage.

- Software Updates: Duplicated software platforms need to be updated periodically to fix bugs, add new features, and improve security. Software updates also provide software compatibility with new technology and information sources. Not updating software leads to slow performance and potential security threats.
- New technology integration: Upon the introduction of new technology, digital twins would then integrate new technologies, like improved sensors or improved software analysis. It will again require more research funding, development, and implementation of new technology.
- Operational Costs: Digital twin system operation and maintenance entail operational costs like employee training, monitoring of the system, and data management. They are needed for effective system operation and production of helpful reports.

Budgeting the cost of such recurring costs is very important to determine the long-term viability of digital twin projects. Budgeting such costs by the cities and including them in their budget is important to prevent service disruption and ensure digital twin efficiency.

12.3.4.2 Cost-Benefit Analysis

The rationale for investment in a digital twin is based on a fine cost-benefit assessment. The assessment allows city planners to demonstrate the value of digital twin technologies and attract funding by identifying potential return on investment (ROI) potential. The most important components of this assessment are:

- Measuring Benefits: City planners have to quantify the benefits of digital twins in actual numbers. This means trying to estimate the increase in efficiency of operation, lowering of costs, better public services, and utilisation of the resources to the maximum.
- Benchmarking and Case Studies: By studying similar projects, we can get a good idea of how useful and profitable digital twins can be. Successful implementations in other sectors or cities can be analysed to justify the expected benefits and support funding requests.

In essence, surmounting the financial and budgetary challenges of digital twins is a function of prudent planning, fiscal discipline, and a tangible value proposition. Through the containment of high upfront expenses, budgeting for operational costs, and rigorous cost-benefit analysis, cities are capable of facilitating the efficient deployment and sustainable feasibility of digital twin technology in urban development.

12.3.4.3 Suggested Solutions to Economic & Budgetary Limitations

To counter the high cost of adoption of digital twins in IIoT settings, cities can embrace phased implementation, commencing with pilot projects yielding high impact to establish ROI before global roll-outs. Using open-source platforms for digital twins can also lower software costs while still ensuring customisation. Along with this, there is a possibility of a financial risk in public and private partnerships, and outcome-based financing models enable cities to reimburse technology suppliers on the basis of realised performance gains, such as energy efficiency or lowered downtime. Cloud-based digital twin services can reduce infrastructure costs while providing scalability for possible extension in the future.

12.3.5 Ethical and Social Implications

The deployment of digital twins in urban planning introduces several ethical and social considerations that must be addressed to ensure that the technology benefits society as a whole and upholds principles of fairness, transparency, and community engagement.

12.3.5.1 Equity and Access

Equal access to the benefits of digital twins is a moral necessity. Use of digital twins in city planning can result in much better public services and infrastructure. Unless taken very seriously, however, the benefits can disproportionately accrue to specific sections of society and can even exacerbate existing inequalities.
Key Characteristics of Equity and Access:

- Avoiding the Digital Divide: The largest challenge is the digital divide—the divide between the digitally connected and the digitally disconnected. If the digital twin technology is predominantly applied in wealthier or better-served communities, less-served communities will not experience the same level of advancement. In order to avoid this, cities must support inclusion in their efforts of digital twins, investing in a manner that will benefit all neighbourhoods.
- Inclusive Planning: Various groups of individuals must be considered while developing digital twins. For example, transportation, healthcare, and environmental management should be made better for everyone, including the poor or economically backward social classes.
- Affordable Access: The technology must be made affordable and accessible to everyone in terms of cost and networks. Cities can try to make free access available to marginalised groups in order to avoid further inequity.

12.3.5.2 Transparency and Accountability

The ethical aspect of digital twin technology demands transparency and accountability in the collection, use, and utilisation of the data. Trust from the public must be built in the successful implementation of digital twins and is dependent on open and transparent practice.

Core Features of Transparency and Accountability

- Open Communication: Policymakers and city planners need to be transparent regarding how digital twin data is being used and upon what decisions are being made based on such data. Transparency about all the information pertaining to data sources, research methods adopted, and decision-making rationale makes the process transparent and builds public trust.

- Governance Frameworks: Appropriate governance frameworks and ethical principles need to be established to utilise digital twins effectively. The frameworks must provide for how the information is to be dealt with, to whom it is to be revealed, and on what grounds decisions are to be taken. They must contain provisions for monitoring and review of compliance with ethical principles.
- Public Reporting: Regular publication of the outcome and impact of digital twin projects keeps people on their toes. It entails releasing success stories, lessons learnt, and problems encountered to the public. Openness in reporting instils confidence and provides valuable lessons for future projects.

12.3.5.3 Public Participation

Public engagement in planning and executing projects for digital twins is central to their attainment and adoption. Public participation ensures that the technology is attuned to the needs and concerns of individuals.
Most Important Characteristics of Public Involvement:

- Community Engagement: Involving the people in the planning and decision-making process ensures that the priorities of the people are met in the digital twin projects. Public consultations, workshops, and advisory committees with representative public members can be employed to attain this.
- The Feedback Mechanisms: Feedback and opinion forums need to be established where citizens are allowed to give their feedback and suggestions to enhance digital twin projects. These could be online questionnaires, public forums, and interactive forums through which people can give their suggestions and concerns.
- Education and Illustration: Public education on digital twin technologies and the potential benefits promotes understanding and acceptance. Public outreach programmes and publicity campaigns can reduce misunderstanding of the technology and promote active participation.
- Dealing with Concerns: Anticipatory dealing with concerns of the public and incorporating the feedback into planning is needed in order to guarantee that there exists trust within society. It is a matter of dealing with data privacy, security, and the potential impact of digital twins on daily life.

With a high regard for equity and openness, accountability, and public engagement, cities will be better equipped to address the ethical and social aspects of digital twins in urban planning. Integrating them into the planning and execution process will ensure that the potential of digital twins is optimised while upholding the standards of ethics and winning public trust.

12.4 TECHNOLOGICAL ADVANCES SUPPORTING DIGITAL TWINS

12.4.1 Internet of Things (IoT)

IoT is central to the creation and operation of digital twins because it provides the connectivity and data necessary.

12.4.1.1 Sensor Networks

IoT devices and sensors are utilised to gather real-time measurements of the physical world. Sensors monitor various parameters such as temperature, humidity, traffic flow, energy consumption, and many more. The data from the sensors is fed into the digital twin in order to create a realistic and up-to-date replica of the physical world.

12.4.1.2 Connectivity

Seamless connectivity of IoT networks allows for seamless data sharing among the physical objects and their digital counterparts. Wi-Fi, Bluetooth, Zigbee, and cellular networks like 5G facilitate data communication securely at high speeds.

12.4.1.3 Edge Computing

For the processing and analysis of the vast amounts of data generated by IoT devices, edge computing is employed. Edge computing allows data to be processed near its origin point, thereby reducing latency and improving response times. This is particularly critical for applications requiring real-time decision-making and control.

12.4.2 Artificial Intelligence (AI) and Machine Learning (ML)

AI and ML play a significant role with respect to data analysis, forecasting, and optimisation of digital twin performance.

12.4.2.1 Predictive Analytics

Past data as well as real-time data can be analysed by AI and ML algorithms to forecast future trends and behaviour. For instance, in urban planning, traffic movement, power consumption, and environmental impacts can be forecasted by AI so that city planners can make informed decisions.

12.4.2.2 Anomaly Detection

ML models can spot anomalies and irregular patterns in data gathered through digital twins. Early detection of potential issues, for instance, infrastructure breakdowns or security breaches, is critical.

12.4.2.3 Optimisation

Urban planning and administration can be optimised by AI through various features. AI, for instance, can optimise traffic light cycles or traffic flows to decongest traffic, optimise energy supply for efficiency, and allot resources in times of emergencies efficiently.

12.4.3 Big Data and Analytics

Big data technologies enable the storage, processing, and analysis of large data needed for digital twins.

12.4.3.1 Data Storage

Effective data storage and management tools for large datasets are provided by big data platforms, such as Hadoop and Apache Spark. Such platforms are able to handle complex data that is created by digital twins in such a way that all the data required is at your fingertips waiting to be analysed.

12.4.3.2 Data Processing

Advanced data processing techniques, including parallel processing and distributed computing, facilitate efficient processing of big datasets. This functionality is essential in real-time city centre analysis and decision-making.

12.4.3.3 Advanced Analytics

Big data analysis software enables the extraction of useful insights from the data collected. Data mining, clustering, and sentiment analysis are a few of the methods that are employed to identify patterns, correlations, and trends affecting urban planning.

12.4.4 5G and Edge Computing

The integration of edge computing and 5G enhances the performance capacity and functionality of digital twin models.

12.4.4.1 High-Speed Connectivity

5G technology offers high-speed, low-latency connectivity that is crucial for the real-time efficient operation of digital twins. The high-bandwidth 5G networks ensure the seamless transmission of huge data from IoT devices to digital twin platforms.

12.4.4.2 Edge Computing

Edge computing reduces latency and bandwidth use by processing data close to the source. This is particularly required for applications with real-time response needs, for example, autonomous vehicles, smart grids, and emergency networks. Edge computing maintains data privacy because it limits data to be sent to centralised servers.

12.4.4.3 Enhanced Functionalities

Edge computing and 5G enable cutting-edge applications such as virtual reality (VR) and augmented reality (AR), which can be incorporated into digital twins. They can also give interactive immersive encounters in the urban management and planning field globally.

12.4.5 Blockchain Technology

Hence, it offers software that protects and accepts transacting activities of digital twin data in an open way.

12.4.5.1 Data Security

The fact that blockchain is decentralised and cannot be altered means the data of the digital twin is secure and intact. All transactions are stored in a ledger that cannot be tampered with, keeping data breaches and unauthorised uses to a minimum.

12.4.5.2 Transparency and Trust

With audibility in transactions demonstrated and ensured by blockchain, trust is established between stakeholders, particularly in urban planning, where the same enhances cooperation and accountability among different stakeholders, whether government, private, or citizens.

12.4.5.3 Smart Contracts

Smart contracts are self-executing contracts with the terms written into lines of code. The smart contracts can be used to automate most digital twins' processes, for instance, by using them to manage service level agreements in automatised scheduling of maintenance and securing data sharing between permissioned parties.

12.5 PROPOSED FRAMEWORK: SECURE DIGITAL TWINS FOR IIOT ENVIRONMENTS

As digital twins come to speed in IIoT environments, cognitive security becomes critical for safeguarding confidentiality, availability, and integrity of data. To respond to these issues, we suggest a Secure Digital Twin Framework for IIoT in this chapter that maximises real-time monitoring while ensuring lightweight security for resource-limited scenarios.

12.5.1 Framework Components

12.5.1.1 Data Sources Layer

IIoT sensors and devices constantly produce real-time operating data (temperature, vibrations, energy usage, etc.) that are input into the digital twin platform.

12.5.1.2 Edge Processing Layer

Lightweight encryption (e.g., ECC-based encryption) and simple preprocessing occur at the edge to cut down on raw, unsecured transmission.

12.5.1.3 Digital Twin Core

The physical system is simulated virtually in real time, utilising the secured data for visualisation, performance monitoring, and predictive maintenance analysis.

12.5.1.4 Blockchain Integration Layer

A permissioned blockchain (such as Hyperledger Fabric) maintains important data transactions with tamper-evident logs to ensure data integrity and trust without centralised control.

12.5.1.5 AI-based Anomaly Detection Module

ML models that can track the digital twin data stream, detecting anomalies that could lead to cyberattacks or operational malfunctions, allowing for proactive detection and mitigation.

12.5.1.6 User Interface Layer

A dashboard that enables only authorised personnel to view the system status, be alerted, and interact with the digital twin for operational decisions.

The proposed Secure Digital Twin Framework for IIoT environments, highlighting data sources, edge processing, blockchain integration, AI-based anomaly detection, and user interaction layers, is depicted in Figure 12.2.

12.5.2 Framework Benefits

12.5.2.1 Data Integrity

The implementation of blockchain guarantees data records that cannot be changed, this increases the confidence in the real-time data streams of the digital twin.

12.5.2.2 Lightweight Security

Edge encryption and sparse blockchain logging balance security needs against resource limitations of IIoT devices.

12.5.2.3 Cognitive Security

AI frameworks allow intelligent monitoring and early detection of abnormal trends, reducing response times for potential attacks or failures.

FIGURE 12.2 Secure digital twins for IIoT environments framework.

infrastructure expansion and upgrading—across industries such as energy, water, sewer, gas, transport, roads, and communications—while facilitating holistic evaluations of existing infrastructure conditions. As a pilot, the project is set to improve urban planning for both the private and public sectors. An example in that practice is The Monarch on MLK—Campus for Economic Mobility. In October 2023, the City of St. Louis Land Clearance for Redevelopment Authority (LCRA), which employs SLDC, purchased a 150,000-square-foot factory building at 3940 Dr. Martin Luther King Drive and 29 vacant parcels of land immediately surrounding it, covering 15.39 acres. The plan calls for redevelopment into a cutting-edge, transit-served workforce and training centre for manufacturing and other high-demand, high-wage occupations. Supported by SLDC, the campus is designed to serve as a community hub—focusing workforce development programmes, services to small and minority-owned businesses, and a variety of community resources and collaborative projects.

12.6.3 Applications and Benefits

12.6.3.1 Community-Based Urban Renewal

The digital twin assists in planning transformations to the MLK Corridor through the display of concepts such as the Campus for Economic Mobility and the 4theVille Cultural Boulevard. Community members are able to walk through these concepts, view various alternatives, and provide feedback. It opens up the process and invites everyone to collaborate.

12.6.3.2 Equitable and Just Economic Growth

The overarching objective of the project is to restore economic value to the region by converting it into a destination for local companies and cultural tourism. The digital twin assists in determining where businesses can expand and where one can invest, while maintaining the cultural identity of the region.

12.6.3.3 Establishing Trust with the Community

The interactive model is easier to comprehend for residents when there are complicated redevelopment plans. They are able to see things change from the street level, allowing them to provide informed feedback. This is a move from the conventional top-down decision-making and goes for working collaboratively with the community.

12.6.3.4 Fair Policies

With digital simulation of changes, the project accommodates growth that is racially and economically equitable. City planners can forecast danger such as displacement, look at the potential impact of traffic, and ensure enhancements benefit communities that were previously excluded.

12.6.3.5 Education and Awareness

The model is also beneficial for learning. It can be employed by students, city planners, and technology designers to learn about the problems and potential of the city. This makes discussions more innovative and solutions more creative.

12.6.4 Future Directions and Challenges

The MLK Corridor digital twin has several challenges, although it is strong. Real-time data integration is as yet limited because the model has predominantly static planning data. Furthermore, scaling the model citywide will need much investment, stakeholder collaboration, and technical upscaling.

In the future, the project seeks to incorporate real-time sensor data, urban simulations using AI, and augmented/virtual reality interfaces. These features would facilitate dynamic city management, ongoing public engagement, and predictive socio-economic modelling. Additionally, incorporating privacy and security mechanisms into the system will be critical with growing data complexity.

Figure 12.6 outlines the envisioned future expansion of the MLK Corridor Digital Twin, including real-time sensor integration, AI-driven simulations, and immersive AR/VR interfaces.

12.6.5 St. Louis MLK Corridor Digital Twin Analysis

The MLK Corridor digital twin is notable for being both socially responsible in its design and first-people approach. Unlike most smart city proposals that give centrality to infrastructure, this model puts people—especially long-marginalised communities—first in its planning rationale.

But, its existing restrictions on live feeds and integrating with old city systems must be resolved in order to realise its maximum real-time applications. For better security and sustainability, the city can consider blockchain-based transparency for urban transactions, and ML algorithms for detecting risks or gentrification trends. The project emphasises that future generations of digital twins should not only engage

FIGURE 12.6 3D digital twin rendering of the St. Louis Digital Commons redevelopment site showing proposed building layout, parking, and green infrastructure.

with technology but also ethical and inclusive deployment in order to become actual tools for justice-oriented smart cities.

12.6.6 Case Study Conclusion

The St. Louis MLK Corridor Digital Twin showcases how digital twins can be effective tools for inclusive urban renewal. It combines immersive technology with balanced planning objectives to facilitate community empowerment, well-informed decision-making, and future-proof city modelling. The project demonstrates the power of collaboration, open standards, and civic engagement, offering a compelling model for cities seeking to combine smart technology with social change. As cities around the world embark on digital transformation, St. Louis provides a template for doing so with intention and people in mind.

12.7 CONCLUSION

12.7.1 Summary

The concept of a digital twin represents a significant innovation in smart city and urban planning. With virtual replicas of physical assets, digital twins allow for city planners to monitor, analyse, and enhance urban operations in real time. Digital twins can transform several aspects of urban living, from infrastructure management to public safety and emergency services. While there are several advantages, digital twin deployment is faced with some difficulties, such as data management, interoperability, security, and resource limitations. These can only be overcome through collaboration among stakeholders, ongoing technological development, and the establishment of strong frameworks.

This chapter also discussed how digital twins can be used to improve smart city planning and IIoT setups with a focus on the importance of cognitive security integration. It introduced a secure digital twin architecture, which utilises blockchain for data integrity, AI for anomaly detection, and lightweight edge encryption to counter data security issues without compromising real-time performance. The critical evaluation of the 3D City Model of Helsinki reinforced the pragmatic advantages of digital twins while emphasising integrated security features in future deployments. Scalable blockchain frameworks in digital twins, efficient lightweight encryption for IIoT devices, and the integration of cognitive security models with negligible computational overhead are some topics that should be investigated in future research to enhance secure digital twin deployments in smart cities.

The path to reaching the full potential of digital twins in urban planning is filled with issues. Data integration and management: the volume and variety of data needed stand in the way of successfully developing digital twins for smart city planning; this process is definitely not easy.

12.7.2 Urban Planners' Recommendations

Urban planners should put more emphasis on building strong data management and integration frameworks to maximise the potential of digital twins. This includes adopting standardised formats for data, protocols, and interoperability solutions to facilitate seamless data transfer from one system to another. Incorporation of sophisticated data analytics and processing equipment has to be invested in to cope with the huge amounts of data generated by digital twins and derive valuable insights useful in decision-making. Cybersecurity practices must be enforced by city planners to safeguard the data collected and processed by digital twins. This means involving security audits, encryptions, and security protocols for safer transfer of data without interception and manipulation. Data anonymisation methods and compliance with privacy laws can provide assurances regarding privacy issues and public confidence. Appropriating financial funding and resources for digital twin projects is important for their successful realisation. Urban planners must conduct cost-benefit analysis to be in a position to justify the investments that they plan for digital twin technology. This will help in securing stakeholder support as well as access to required finance.

12.7.3 Future Research Directions

The future work in the domain of digital twins has to overcome the existing challenges and create new applications and innovations.

According to the framework suggested in this chapter, the following research questions are suggested:

- In what ways can blockchain scalability be maximised in high-speed IIoT data streams without latency in digital twin systems?
- Which lightweight cryptographic mechanisms can be incorporated within edge processing to protect IIoT data with little computational burden?
- How should AI cognitive security models be properly trained on the limited datasets characteristic in certain industrial settings?

One such piece of research in this regard is creating higher-level data integration and interoperability solutions. This involves designing data formats, protocols, and interfaces to be standardised in such a way that data can be shared among systems and technologies of varying types in an effective way. Enhancing cybersecurity and privacy controls for digital twins is a significant research field too. Advanced encryption methods, secure communication protocols, and strong authentication mechanisms can be developed to safeguard the data gathered and processed by digital twins. Additionally, anonymisation techniques research and privacy-preserving data analytics research can be performed to minimise privacy concerns and maintain compliance with the law. Future research also needs to consider how ethical digital twin application development principles and frameworks can be made realistic. This includes the design of data governance, openness, and accountability frameworks and how digital twin projects can be organised to satisfy societal values and priorities. Interaction with a diverse range of stakeholders, including communities, policymakers, and industry stakeholders, will determine the ethical context of digital twin technology.

BIBLIOGRAPHY

1. Batty, M. (2018). Digital Twins. *Environment and Planning B: Urban Analytics and City Science*, 45(5), 817–820. https://doi.org/10.1177/2399808318796416
2. Ketzler, B., Nunes, V., Lobo, J. P., Oliveira, A., Simões, M., & Silva, A. (2020). Digital Twins for Smart Cities: A Literature Review. *Built Environment*, 46(4), 547–573. https://doi.org/10.2148/benv.46.4.547
3. Lu, Q., Xie, X., Heaton, J., Parlikad, T., & Schooling, J. (2020). Digital Twins in Civil Engineering: A State-of-the-Art Review. *Journal of Management in Engineering*, 36(3), 04020002. https://doi.org/10.1061/(ASCE)ME.1943-5479.0000760
4. Grieves, M., & Vickers, J. (2017). Digital Twin: Mitigating Unpredictable, Undesirable Emergent Behaviour in Complex Systems. In Franz-Josef Kahlen, Shannon Flumerfelt, & Anabela Alves (Eds.), *Transdisciplinary Perspectives on Complex Systems* (pp. 85–113). Springer. https://doi.org/10.1007/978-3-319-38756-7_4
5. European Commission (2020). Digital Twins: Bridging Physical and Digital for Smart Cities. https://ec.europa.eu/digital-single-market/en/news/digital-twins
6. Batty, M., Axhausen, K. W., Giannotti, G., Pozdnoukhov, A., Bazzani, A., Wachowicz, M., Ouzounis, G., & Portugali, Y. (2012). Smart Cities of the Future. *The European Physical Journal Special Topics*, 214(1), 481–518. https://doi.org/10.1140/epjst/e2012-01703-3
7. Hollands, R. G. (2008). Will the Real Smart City Please Stand Up? *City*, 12(3), 303–320. https://doi.org/10.1080/13604810802479126
8. Allam, Z., & Newman, P. (2018). Redefining the Smart City: Culture, Metabolism, and Governance. *Smart Cities*, 1(1), 4–25. https://doi.org/10.3390/smartcities 1010002
9. Albino, V., Berardi, U., & Dangelico, R. M. (2015). Smart Cities: Definitions, Dimensions, Performance, and Initiatives. *Journal of Urban Technology*, 22(1), 3–21. https://doi.org/10.1080/10630732.2014.942092
10. McKinsey Global Institute (2018). Smart Cities: Digital Solutions for a More Liveable Future. www.mckinsey.com/industries/capital-projects-and-infrastructure/our-insights/smart-cities
11. Zanella, A., Bui, N., Castellani, A., Vangelista, L., & Zorzi, M. (2014). Internet of Things for Smart Cities. *IEEE Internet of Things Journal*, 1(1), 22–32. https://doi.org/10.1109/JIOT.2014.2306328
12. Tao, F., Zhang, H., Liu, A., & Nee, A. Y. C. (2019). Digital Twin Driven Smart Manufacturing: Connotation, Reference Model, Applications, and Research Issues. *Robotics and Computer-Integrated Manufacturing*, 61, 101837. https://doi.org/10.1016/j.rcim.2019.101837
13. Jha, S., Jha, N., Prashar, D., Ahmad, S., Alouffi, B., & Alharbi, A. (2022). "Integrated IoT-Based Secure and Efficient Key Management Framework Using Hashgraphs for Autonomous Vehicles to Ensure Road Safety," *Sensors*, 22(7), 2529. https://doi.org/10.3390/s22072529
14. Cisco (2020). 5G in Smart Cities. www.cisco.com/c/en/us/solutions/industries/smart-connected-communities.html
15. Hashem, I. A. T., Yaqoob, I., Anuar, N. B., Mokhtar, S., Gani, A., & Khan, S. U. (2016). The Role of Big Data in Smart Cities. *International Journal of Information Management*, 36(5), 748–758. https://doi.org/10.1016/j.ijinfomgt.2016.05.002
16. Roman, R., Zhou, J., & Lopez, J. (2013). On the features and challenges of security and privacy in distributed internet of things. *Computer Networks*, 57(10), 2266–2279. https://doi.org/10.1016/j.comnet.2013.02.011

17. Weber, R. H. (2010). Internet of Things—New Security and Privacy Challenges. *Computer Law & Security Review*, 26(1), 23–30. https://doi.org/10.1016/j.clsr.2009.11.008
18. NIST (2020). Privacy Framework: A Tool for Improving Privacy through Enterprise Risk Management. www.nist.gov/privacy-framework
19. Jha, S., Nkenyereye, L., Joshi, G. P., & Yang, E. (2020). "Mitigating and Monitoring Smart City Using Internet of Things," *CMC-Computers, Materials & Continua*, 65(2), 1059–1079. https://doi.org/10.32604/cmc.2020.011754
20. Allam, Z., & Dhunny, Z. A. (2019). On Big Data, Artificial Intelligence, and Smart Cities. *Cities*, 89, 80–91. https://doi.org/10.1016/j.cities.2019.01.032
21. Jha, S., Ahmad, S., Abdeljaber, H. A. M., Othman, N. A., Nazeer, J., Yadav, K., & Alkhayyat, A. H. (2024). "Enabling Resilient Wireless Networks: OFDMA-Based Algorithm for Enhanced Survivability and Privacy in 6G IoT Environments," *IEEE Transactions on Consumer Electronics*.https://doi.org/10.1109/TCE.2024.3370414
22. Future Cities Catapult (2017). The Role of Digital Twins in Smart Cities. https://futurecities.catapult.org.uk/2017/11/15/digital-twins-smart-cities/
23. Zhang, H., Jin, C., Cui, H. (2018). A Method to Predict the Performance and Storage of Executing Contract for Ethereum Consortium-Blockchain. In Chen, S., Wang, H., Zhang, LJ. (Eds.), *Blockchain – ICBC 2018. Lecture Notes in Computer Science*, vol. 10974. Springer. https://doi.org/10.1007/978-3-319-94478-4_5
24. Jha, S., Ahmad, S., Arya, A., Alouffi, B., Alharbi, A., Alharbi, M., & Singh, S. (2023). "Ensemble learning-based hybrid segmentation of mammographic images for breast cancer risk prediction using fuzzy c-means and CNN model," *Journal of Healthcare Engineering*, vol. 2023.https://doi.org/10.1155/2023/1491955
25. Siemens (2020). Digital Twin City Case Studies. https://new.siemens.com/global/en/company/topic-areas/digital-twin.html
26. Premsankar, G., Di Francesco, M., & Taleb, T. (2018). Edge Computing Enabled Smart Cities: A Case Study. *IEEE Internet of Things Journal*, 5(2), 1275–1284. https://doi.org/10.1109/JIOT.2018.2805263
27. Khan, Z., Anjum, A., Kiani, S. L., & Hussain, A. J. (2020). Digital Twin Framework for Smart Mobility in Cities. *Sensors*, 20(18), 5271. www.mdpi.com/1424-8220/20/18/5271
28. Neirotti, P., De Marco, A., Cagliano, A., Mangano, G., & Scorrano, F. (2014). Current Trends in Smart City Initiatives: Some Stylised Facts. *Cities*, 38, 25–36. https://doi.org/10.1016/j.cities.2013.12.010
29. Zhang, Y., Yu, X., Xie, R., & Qiu, G. (2020). Blockchain for Smart City: A Survey. *IEEE Communications Surveys & Tutorials*, 23(2), 902–943. https://doi.org/10.1109/COMST.2019.2899617
30. Bibri, S. E. (2021). The Internet of Things for Smart and Sustainable Cities. *Computers, Environment and Urban Systems*, 85, 101584. https://doi.org/10.1016/j.compenvurbsys.2020.101584
31. Jha, S., Kumar, R., Son, L. H, Abdel-Basset, M., Priyadarshini, I., Sharma, R., & Lon, H. V. (2019). Deep learning approach for software maintainability metrics prediction. *IEEE Access, 7*, 61840–61855. https://doi.org/10.1109/ACCESS.2019.2913349
32. Hassanien, A. E., & Awad, A. I. (2020). Digital Twin for Smart Cities: A Survey. *Internet of Things*, 12, 100313. https://doi.org/10.1016/j.iot.2020.100313
33. St. Louis Development Corporation. (2025). SLDC Commissions Digital Twin of MLK Corridor, Unveils Name of Campus for Economic Mobility. www.developstlouis.org/news-media/sldc-commissions-digital-twin-of-mlk-corridor-unveils-name-of-campus-for-economic-mobility

34. Minerva, R., Lee, G. M., & Crespi, N. (2020). Digital Twin in the IoT Context: A Survey. *Proceedings of the IEEE*, 108(10), 1785–1824. https://doi.org/10.1109/JPROC.2020.2998530
35. Jones, D., Snider, C., Nassehi, A., Yon, J., & Hicks, B. (2020). Characterising the Digital Twin: A Systematic Literature Review. *CIRP Journal of Manufacturing Science and Technology*, 29, 36–52. https://doi.org/10.1016/j.cirpj.2020.02.002
36. Lee, E. A.. (2008). "Cyber Physical Systems: Design Challenges," Tech. Rep. UCB/EECS-2008-8, EECS Dept., Univ. of California, Berkeley, CA, USA. [Online]. http://www2.eecs.berkeley.edu/Pubs/TechRpts/2008/EECS-2008-8.html

13 Case Studies and Applications of Cognitive Security in Industry

Jackulin Mahariba A

13.1 INTRODUCTION

Cognitive Security is a comprehensive Security strategy combining Artificial intelligence (AI), Machine Learning (ML) along with Cognitive Computing solutions for the detection and prevention of threats and response to events. Cognitive security systems analyze large amounts of structured and unstructured data, detect abnormal behaviors called as "anomalies" and predict future threats by simulating human-like reasoning and learning capabilities.

The key elements of cognitive security [1] are defined as follows:

AI Integration: Automates detection of anomalies and cyber threats by processing data at speeds far beyond human capabilities.

Machine Learning: Learns from new data and adapts its threat-detection algorithms to new cybersecurity threats.

Cognitive Computing: Simulates human reasoning by interpreting natural language, analyzing context, and providing actionable insights.

With these capabilities, cognitive security strengthens an organization's ability to defend against known vulnerabilities as well as proactively detecting and mitigating emerging threats. It is increasingly critical in modern cybersecurity frameworks to tackle the precision [2], scalability, and adaptability in response to today's cyberattacks.

In our digital age where threats from every side are becoming increasingly complex and sophisticated, cognitive security has emerged as a core technology for modern industry [3–5]. As industries become digitally connected and their systems become interconnected, traditional security measures can no longer protect against advanced persistent threats (ATP). Cyberattacks are becoming more sophisticated, using AI and automation to exploit these holes. Threat actors employ zero-day exploits, ransomware, and deepfake technologies that require equally sophisticated defensive mechanisms. Industries also produce enormous amounts of data every day that manual or rule-based systems cannot process and analyze in real-time. Cognitive security systems can cope with this data deluge and provide fast, accurate threat analysis, and response. The proliferation of IoT devices has widened the attack surface

DOI: 10.1201/9781003609773-13

and added new holes. Cognitive security helps in detecting and mitigating threats in such large, distributed networks. Unlike reactive security, cognitive security detects potential vulnerabilities and mitigates risks before they occur. ML models adapt to new attack vectors [6, 7] and defend against them.

Cognitive security is needed to protect critical infrastructure, financial systems, health data, and intellectual property in industries such as [8, 9]:

- Finance: Detects fraud and secures financial transactions.
- Healthcare: Protects sensitive patient data and connected medical devices.
- Manufacturing: Secure industrial control systems and supply chains.
- Retail: Limits risk in e-commerce and protects customer information.

With intelligent, adaptive, and scalable solutions, cognitive security fulfills the need for robust cybersecurity in an era of digital transformation and hyper-connectivity. It enables industries to keep pace with adversaries and to adapt to emerging threats.

The main aim of this chapter is to describe and analyze how cognitive security can change modern industries through real-world applications and case studies. Through the lens of practical case studies and applications, this chapter aims to demonstrate how cognitive security is transforming the cybersecurity landscape and driving resilience in the face of growing digital threats [10, 11].

13.2 IMPACT OF COGNITIVE SECURITY IN INDUSTRIES

Cognitive security addresses the increasing sophistication of cyber threats through AI, ML, and cognitive computing. These technologies allow companies to safeguard their systems and data intelligently and proactively, making cybersecurity much more efficient and precise than ever before.

AI and ML are among the core components of cognitive security. Instead of traditional systems deploying static rules and signature-based methods, cognitive security systems employ ML algorithms to analyze massive amounts of data, detect trends and identify emerging threats in real time. As an example, in financial services, ML models interpret transaction information to detect suspicious patterns which could be indicative of fraud. These models continue to learn from new attacks to improve their predictive abilities.

Another critical facet of cognitive security is Natural Language Processing (NLP) in threat intelligence. NLP allows systems to process unstructured data like emails, social media posts, and online accounts and derive actionable insights. For example, cognitive security platforms can scan employee emails for phishing attempts by analyzing the language for signs of malicious intent or urgency, which characterize phishing scams. NLP can also monitor global threat intelligence reports and identify emerging cyber risks to help organizations stay ahead of attacks.

Cognitive security systems are further enhanced by behavioral analysis and anomaly detection. These systems monitor baseline behaviors for users, devices, and networks. Any deviation from this baseline is logged as an anomaly and may indicate a security breach. For example, if an employee's account starts accessing sensitive data at irregular hours or from remote locations, cognitive security systems can notify administrators about the activity and even block it until it is verified. Such capabilities are of particular importance for detecting insider threats as well as APT avoiding conventional defenses.

A key differentiator between cognitive security and traditional security is their approach to threat management. Conventional systems are generally reactive, responding to threats only after they are identified or after damage has occurred. On the other hand, cognitive security is proactive. It identifies threats and predicts vulnerabilities based on historical data and potential attack patterns. For example, in manufacturing, cognitive systems can analyze industrial IoT (IIoT) device logs to predict and limit cyber risks that could down production lines.

In addition, cognitive security provides automated response capabilities. Traditional systems typically require human intervention to analyze and respond to threats, which can delay mitigation efforts. However, cognitive systems do this automatically and respond to threats in real time. For example, a cognitive security system that detects ransomware in a corporate network can isolate infected machines, stop malware spread, and even start recovery protocols without human intervention.

The benefits of cognitive security are transformative. The most important advantage is the real=time threat detection. In industries like healthcare where patient data needs to be protected, cognitive security systems can detect and neutralize threats before they compromise sensitive information. Another important benefit is adaptability and scalability. These systems can deal with huge quantities of data and can adjust to dynamic threat vectors, ideal for dynamic environments such as cloud-based environments.

Finally, cognitive security reduces false positives, which are common with traditional security systems. False alarms waste time and resources because they require unnecessary investigations. AI-powered cognitive systems alleviate this issue by providing more accurate threat detection. For example, in the retail sector, such systems could distinguish between malicious attempts to gain access to customer data and benign anomalies due to routine operations, thus allowing security teams to focus on actual threats only.

Finally, cognitive security rewrites the cybersecurity paradigm by combining AI, ML, NLP, and behavioral analytics to build intelligent, proactive and adaptive defense mechanisms. Its ability to identify threats in real time, anticipate new challenges and reduce false positives makes it indispensable in a digital/connected world. Whether it concerns the protection of financial transactions, healthcare data, or industrial control systems, cognitive security provides a robust and scalable approach to modern cyber threats, ensuring resilience and trust in critical systems.

13.3 APPLICATIONS OF COGNITIVE SECURITY IN INDUSTRY

In a digital world, cybersecurity threats are constantly evolving and require intelligent and adaptive solutions for industries covering financial sector, healthcare, industries involving manufacturing retail, energy, and utilities. Integrated with advanced technologies, cognitive security provides real-time threat detection, response, and prevention which is invaluable in many sectors.

13.3.1 Financial Sector

Cognitive security has benefited the financial industry which is a target for cybercriminals. One important application is real-time fraud detection. Traditional systems typically use static rules that may not be able to detect sophisticated fraud schemes. Cognitive security systems, however, analyze massive amounts of transaction data and find anomalies indicative of fraud such as sudden large withdrawals or unusual purchases. For example, banks use ML models to track customer behavior and flag transactions straying from standard usage so they can take action to avoid losses.

Another important application is to secure online transactions. With the move toward digital payment systems, financial institutions are also at risk for phishing and transaction tampering. Cognitive security systems improve transaction security by authenticating user behavior, detecting malicious activities, and encrypting communications between parties using AI. For instance, NLP can analyze phishing emails from financial clients and block them before the user opens the email.

13.3.2 Healthcare

The healthcare sector is facing unique cybersecurity challenges with the rise of smart hospitals and other connected devices. Patient data should be protected from breaches which can result in serious privacy violations and even compromise patient safety. Cognitive security systems protect electronic health records (EHRs) by monitoring access patterns, detecting unauthorized attempts, and encrypting sensitive information. For instance, AI-driven systems can control access to patient data by healthcare staff to avoid data leakage in violation of privacy regulations.

Moreover, cognitive security is critical for the defence against ransomware attacks on medical devices. As connected devices such as infusion pumps and imaging systems become more common, they introduce new vulnerabilities. Cognitive systems can detect unusual device behavior, isolate compromised equipment, and restore functionality with minimal downtime. In one case, a hospital used cognitive security to deflect a ransomware attack on its imaging systems without interruption of patient care.

13.3.3 Manufacturing

Manufacturing industries have adopted IoT devices and smart factories that increase productivity but increase cybersecurity risks. Cognitive security controls IoT devices to monitor their activity, detect vulnerabilities, and prevent unauthorized access. For example, in an automotive factory, AI-driven security systems monitor robotic arms and assembly lines for abnormal behavior and prevent them from running.

Another important application is the prevention of industrial espionage. Often manufacturing firms are threatened by competitors that want to steal intellectual property. Cognitive security systems analyze network traffic, detect data exfiltration attempts, and block unauthorized transmissions. For example, a semiconductor manufacturer used cognitive security to defend its proprietary chip designs against cyber intrusions.

13.3.4 Retail

The retail sector and especially the e-commerce domain rely on cognitive security to maintain customer trust and secure operations. It's a top priority to safeguard e-commerce platforms which are frequently targets of payment fraud and data leaks. Cognitive systems analyze purchase behavior, detect fraudulent transactions and secure payment gateways in real time. For instance, an internet store used ML algorithms to detect fake accounts used for fraudulent activities.

Cognitive security also provides secure experiences that increase customer trust. Protecting sensitive customer data such as payment information and personal details, cognitive systems enable a secure shopping experience. Retailers that deploy such systems show they care about cybersecurity, and build relationships with their customers.

13.3.5 Energy and Utilities

Cyber-physical threats are unique to the energy and utilities sector where critical infrastructure is at risk. Cognitive security is vital for monitoring and protecting critical infrastructure as power lines and water treatment facilities. AI-powered systems analyze sensor data of these infrastructures to detect irregularities such as unauthorized access or operational anomalies and take immediate action to mitigate risks.

Cognitive security is also required to detect and limit cyber-physical attacks with devastating results. For example, a natural gas pipeline operator used cognitive security systems to identify unusual pressure changes caused by a cyberattack, preventing a potential explosion. These systems ensure the resilience of essential services and protect against disruptions with potentially large societal impacts. Cognitive security has proven itself to be a game-changer across industries providing intelligent, proactive ways to meet today's cybersecurity threats. Whether it is detecting fraud in financial transactions, securing patient data in healthcare, securing IoT devices in manufacturing, securing e-commerce platforms or defending critical infrastructure in the energy sector, cognitive security enables industries to operate safely and securely. As cyber threats evolve, cognitive security will become a prerequisite for a resilient and trustworthy world.

13.4 CASE STUDIES

To show the impact of cognitive security and its benefits in the industrial sector, case studies in three domains are explored and presented in this section. The predominant

and middleware that enable communication and data exchange with these legacy systems. For example, a bank might combine its new fraud detection module with an older transaction monitoring system. The AI system analyzes incoming transactions in real time, flags suspicious activities and reports this to the legacy system for further action. This approach keeps existing workflows functional while augmenting them with AIs advanced capabilities. Tools such as Mulesoft or Apache Camel can provide the middleware to enable such integrations and deployment with minimal disruption.

Together, data lakes, AI pipelines and integration with legacy systems provide an architecture to enable the deployment of advanced solutions such as fraud detection. This architecture, which centralizes data storage, automates learning processes and is compatible with existing infrastructure, gives banks and cybersecurity the scalability, adaptability and efficiency they require to face modern challenges. The benefits of using above-discussed technical architecture results in the following outcomes.

- The first year of implementation of the AI-driven threat detection system showed 80% reduction in fraud cases and reduction in financial losses.
- The system neutralized phishing attempts with over 95% accuracy, preventing customer data breaches.
- Customer trust improved, as evidenced by positive feedback and increased use of the bank's online services.

13.4.2 Healthcare Data Breach Prevention

The increasing digitization of healthcare systems has several advantages including better patient care and medical record control. However, this progress has also created new vulnerabilities in connected systems that store and process sensitive patient data. One healthcare organization needed to defend its networked infrastructure from cyberattacks while adhering to strict data protection regulations. This major problem in the digitalization of healthcare systems exposes patient data vulnerability in connected systems.

The healthcare organization maintained a network of connected systems including an EHR, diagnostic devices, and telemedicine platforms. These systems, though integral to modern healthcare delivery, were particularly vulnerable to cyberattacks. Threats included unauthorized access, ransomware attacks, and data theft. Complicating matters was the challenge of managing security across a heterogeneous IT environment that included legacy systems with limited security features. This exposed patient data to possible breaches, which could compromise patient privacy and the organization's reputation.

The afore-mentioned problem of patient data vulnerability requires an automated Cognitive Security Solutions through Real-Time Monitoring and Encryption. To address these issues, the organization implemented a cognitive security framework. This solution applied AI and ML to perform real-time monitoring, anomaly detection and predictive threat analysis. AI-powered tools continuously analyzed network traffic and user behaviors to detect and respond to potential threats before they escalated.

Strong encryption protocols for encrypting data at rest and in transit were also introduced. End-to-end encryption kept private patient information secret from unauthorized parties – even if it was intercepted. For cognitive security, the organization deployed centralized threat intelligence platforms that pulled together data from all connected systems to enable early detection and response to emerging threats.

The solution applied a cognitive security framework combining AI, ML, and encryption protocols to provide real-time monitoring, proactive threat detection and data protection. The technical breakdown of the implementation of cognitive security framework at various instances in the healthcare domain together with real-world example is provided in the following subsections.

13.4.2.1 Real-Time Health Monitoring Using AI and ML

AI-powered systems monitored network traffic, user activities, and device behaviors across the connected systems of the healthcare organization. They consisted of EHR servers, diagnostic imaging equipment, and telemedicine platforms.

AI models established a baseline of normal network behavior based on historical data. For example, it learned patterns like typical login times, frequently accessed files and common data transfer sizes. Any deviation from baseline (for example, a user downloading an unusually large volume of patient data or accessing records at odd hours) triggered alerts. When anomalies were detected, the system automatically initiated countermeasures such as isolating affected devices or blocking suspicious IP addresses.

Real-Time Example

A hospital detects suspicious login attempts from an overseas IP address targeting its EHR system. AI flags this as a credential-stuffing attack and blocks the IP address to prevent unauthorized access.

13.4.2.2 Predictive Threat Analysis

ML models were trained to predict future threats based on past incidents and emerging cyberattack trends. Combining historical breach data with external threat intelligence feeds, the system identified high-risk scenarios and prepared preventive measures.

Threat intelligence feeds from sources such as the "MITRE ATT&CK" framework [2] provided real time information on new malware strains, phishing techniques, and vulnerability exploits. This framework generates matrices based on the domain of application and identifies the threats. The system assigned *Risk scores* to all connected devices and users based on outdated software, abnormal activity, or external threat indicators. The updates for high-risk systems were prioritized and additional monitoring layers such as Multi-Factor Authentication (MFA) were applied.

Real-Time Example

It does this by flagging suspicious links in emails and sending them to quarantine for review.

13.4.2.3 Encryption for Data Security

End-to-end encryption was used to protect patient data at every point – in databases, between systems, or accessible by authorized personnel.

Patient data stored in EHR systems was encrypted using Advanced Encryption Standard (AES-256) so that data could not be accessed illegitimately without the encryption key. Secure Sockets Layer (SSL) / Transport Layer Security (TLS) protocols were employed to encrypt data exchanges between devices and systems such as lab results to doctors or diagnostic images. The encryption keys were managed via a Hardware Security Module (HSM) to prevent unauthorized decryption.

Real-Time Example

A radiology department sends a patient's Magnetic resonance imaging (MRI) scan to a specialist in another city. The data is encrypted during transmission using TLS to stop it from being intercepted or modified by crooks.

13.4.2.4 Centralized Threat Intelligence Platform

A centralized platform aggregated data from all systems to give an overall view of the organization's security posture. This allowed the security team to track threats, monitor responses, and create compliance reports.

APIs were developed to connect legacy systems, diagnostic devices, and modern platforms with data entering the central system. Security dashboards showed real-time metrics like number of threats detected, blocked attacks, and system vulnerabilities. When a potential breach was detected, the platform enabled cross-system communication to quickly contain and mitigate the threat.

Real-Time Example

During a ransomware attack, the centralized system isolates the infected server from the network and performs an automatic backup recovery to reduce downtime and loss of data.

13.4.3 Securing IIoT

IIoT has enabled smart factories with connected devices and systems to improve operational efficiency in manufacturing. However, this interconnectedness has also created significant cybersecurity challenges. This case study investigates how behavioral analytics was applied to address network vulnerabilities in a smart factory and how it reduced downtime due to cyberattacks.

The major problem in factories having IIoT is network vulnerabilities in a smart factory ecosystem. One of the leading manufacturing companies deployed a smart factory setup with connected IoT devices, sensors, and control systems to optimize production and maintenance. These devices collected and sent data over the network to monitor equipment performance, track inventory, and control production lines in real time. This infrastructure improved efficiency but it also exposed the factory to network vulnerabilities. The potential causes are,

- **Unauthorized access:** IoT devices lacked robust authentication mechanisms and thus were susceptible to unauthorized access and manipulation.
- **Insecure Communication Protocols**: Unencrypted data transmission over the network increased the risk of interception and tampering.
- **Zero-Day vulnerabilities**: New, undiscovered vulnerabilities in IoT devices and systems were exploited by attackers.

- **Insufficient Visibility**: The sheer number of devices prevented timely monitoring of network activities and detection of potential threats.

This led to frequent disruptions such as system outage and equipment downtime, which affected productivity and led to significant financial losses for the factory. To meet these challenges, the company deployed behavioral analytics to detect anomalies and secure the IIoT infrastructure. Behavioral analytics applies advanced algorithms and ML to monitor, analyze, and identify normal behavior in the network. The procedure to execute behavioral analytics in a smart factory is explained using the below steps.

1. **Data Collection:** Data was collected from all connected IoT devices, sensors, and control systems. This included network traffic, device activity logs, and user interactions.
2. **Baseline Behavior Modeling:** ML models were trained to establish a baseline of normal behavior for each device and system. For instance, a certain sensor's typical data transmission patterns or a machine's usual operating parameters.
3. **Real Time Anomaly Detection**: The system continuously monitored network activity compared with the baseline. Any deviation, such as unusual spike in data transmission or unauthorized device communication was flagged as an anomaly.
4. **Automated Threat Response (ATR):** Anomalies detected by the system triggered automated responses such as isolating the device, blocking malicious IP addresses, or informing the security team to investigate further.

This solution was integrated with the factory IT infrastructure and operational technology (OT) systems, causing minimal downtime during deployment.

Real-Time Example

In a cyberattack incident, the behavioral analytics system detected an IoT sensor trying to communicate with an external server outside its normal operation scope. This was flagged as an anomaly. The system quarantined the sensor and alerted the security staff of a potential data breach.

The major outcome of significant decrease in downtime due to cyberattacks were:

1. Behavioral analytics enabled the factory to improve its cybersecurity posture.
2. Production schedules were maintained and financial losses due to disruptions were minimized.
3. Security teams were able to see into the network more effectively to monitor device activity and respond to threats.
4. The success of behavioral analytics solution increased confidence in the smart factory's IIoT infrastructure and investment in automation and connectivity.
5. The solution met industry standards for cybersecurity like International Organization for Standardization (ISO) / International Electrotechnical Commission (IEC) 27001 and NIST guidelines, thereby enhancing the company's image as a secure and reliable manufacturer.

This case study illustrates how behavioral analytics can be applied to address network vulnerabilities in an IIoT-enabled smart factory. Using advanced anomaly detection and ATR, the manufacturing company was able to mitigate cybersecurity risks and reduce downtime. This proactive approach underscores the need to incorporate state-of-the-art security technologies into IIoT ecosystems to preserve operational continuity and resilience to evolving threats.

13.5 CHALLENGES IN ADOPTING COGNITIVE SECURITY

Cognitive security has become critical in the modern era of increasing frequency and sophistication of cyber threats. Cognitive security employs AI, ML, and data analytics to detect and thwart these risks in real time. While the potential of cognitive security is enormous, several barriers prevent its wide deployment. This chapter explores four key challenges: High initial costs, data privacy concerns, false positives, and resistance to change.

13.5.1 High Initial Cost

The large initial investment is perhaps the biggest barrier to adopting cognitive security solutions. Organizations must invest heavily in the acquisition and deployment of AI-driven technologies (hardware, software, and infrastructure upgrades). They also require periodic updates and maintenance to withstand changing threats. Cognitive security solutions may often require computational power to analyze massive amounts of data in real time. This requires investments in high performance servers, Graphics Processing Units (GPUs) and cloud services. These expenses can be prohibitive for Small and Medium-sized Enterprises (SMEs), which cannot be competitive with bigger corporations which can afford such technologies.

Implementing and managing cognitive security systems require professionals like data scientists, cybersecurity experts and AI specialists. But demand for such experts worldwide is much higher than supply, driving up salaries. Organizations must invest in hiring such specialists and continuing training to keep them informed of developments. Many cognitive security solutions are offered by third parties and their licensing models are expensive. Besides, businesses might have to buy extra tools and integrate them with existing systems. Another big hurdle is Return on Investment (ROI) uncertainty. Cognitive security can prevent costly breaches, but the benefits are often intangible and long-term and the upfront investment is often not justified for organizations.

13.5.2 Data Privacy Concerns: Handling Sensitive Data While Ensuring Compliance

Cognitive security systems operate on enormous amounts of data. But the collection, storage and processing of sensitive information create significant privacy and compliance issues. Organizations must adhere to Data Protection Regulations like the General Data Protection Regulation (GDPR) in Europe and also the California Consumer Privacy Act (CCPA) in the United States. These laws specify how data might be collected, stored and utilized. Non-compliance can lead to heavy fines and reputational harm, so organizations are cautious of implementing data-intensive cognitive security solutions. Cognitive security systems often analyze personal data, financial records, and proprietary business data. Confidentiality of this information is critical as a breach or misuse could be disastrous to individuals or organizations.

AI in security poses concerns about who has permission to access the information and just how it's used. Organizations must address these trust issues through effective data governance policies and transparency in their operations. Managing cross-border data transfers while complying with local regulations is another layer of complexity.

13.5.3 False Positives: Challenges in Fine-Tuning Algorithms

Cognitive security systems recognize threats through algorithms. These algorithms are not perfect, and they sometimes produce false positives (when the system incorrectly marks benign activities as malicious). False positives may cause unnecessary alerts and actions (operational disruptions). For example, a false positive – based blocking of a legitimate user's access to critical systems can impede productivity and frustration. Investigating false positives takes time and resources. Security teams must sort through many alerts to find real threats – leading to fatigue and lower efficiency.

The procedure for fine-tuning algorithms to minimize false positives is complicated and ongoing. It demands deep understanding of AI and cybersecurity as well as regular monitoring and adjustments in response to new data and emerging threats. Frequent false positives can undermine trust in the cognitive security system. Personnel and stakeholders may question its effectiveness and be less likely to rely on it or follow its recommendations. To tackle this challenge, organizations can use advanced techniques like anomaly detection, behavioral analysis, and contextual understanding. But such solutions often require additional investments and expertise, adding to the high initial costs.

13.5.4 Resistance to Change: Legacy Systems and Human Factors

Adopting cognitive security solutions often requires significant changes to an organization's existing systems and workflows. Resistance to these changes can come from both technical and human factors.

Many organizations rely on legacy systems that were not designed to work with modern cognitive security solutions. Integrating these systems can be challenging and may require extensive customization. In some cases, organizations may need to replace outdated infrastructure entirely, which can be both costly and time-consuming. Employees and decision-makers may not fully understand the benefits of cognitive security, leading to skepticism and reluctance to adopt it. Misinformation and fear of AI replacing jobs can further exacerbate resistance.

The adoption of cognitive security often requires a cultural shift within the organization. Employees must be willing to embrace recent technologies and workflows, which may involve unlearning old habits and acquiring new skills. Implementing cognitive security solutions involves more than just technological changes; it also requires effective change management strategies. Organizations must communicate the value of these solutions, provide adequate training, and address employee concerns to ensure a smooth transition.

To overcome resistance, organizations can take the following steps:

- **Education and Training:** Conduct workshops and training sessions to familiarize employees with cognitive security and its benefits.
- **Stakeholder Engagement:** Involve stakeholders at all levels in the planning and implementation process to build consensus and address concerns.
- **Incremental Implementation:** Adopt a phased approach to implementation, starting with pilot projects to demonstrate value and build confidence.

While cognitive security holds the promise of revolutionizing cybersecurity, its adoption is fraught with challenges. High initial costs, data privacy concerns, false positives, and resistance to change are significant barriers that organizations must address to fully realize its potential. By investing in education, fostering collaboration, and leveraging advanced technologies, organizations can overcome these challenges and build a robust cybersecurity framework that is capable of tackling the threats of the digital age.

13.6 FUTURE TRENDS IN COGNITIVE SECURITY

Cognitive security is a rapidly developing area which applies AI, ML, and data analytics to defend systems from growing cyber threats. As cybercriminals continue to leverage more sophisticated methods, technological advancements and industry collaboration will define the future of cognitive security. Four trends that will shape the future of cognitive security are examined here: integration with quantum computing, AI explainability, collaborative threat intelligence (CTI), and autonomous systems.

13.6.1 Integration with Quantum Computing

Quantum computing is set to change many areas including cybersecurity. Cognitive security systems also can benefit from the huge computational power of quantum computers which can process quantum amounts of information and solve problems at unpredicted speeds. Conventional encryption methods rely on algorithms that quantum computers can break. But, the combination of quantum cryptography with cognitive security, utilizing principles of quantum mechanics to protect communications can provide unbreakable encryption standards. Cognitive security systems can use quantum cryptographic techniques to defend sensitive data against classical and quantum cyberattacks.

Quantum computing can enable AI models to analyze huge amounts of data in real time. This can help in detecting subtle patterns and anomalies associated with sophisticated cyber threats. Cognitive security systems based on quantum computing can predict and mitigate threats before they occur. However, quantum computing is still in its infancy and will require significant investments in infrastructure and talent to implement in cognitive security. Moreover, organizations need to address issues of interoperability of quantum systems with existing cybersecurity frameworks.

13.6.2 Improved AI Explainability for Compliance and Trust

As AI becomes a cornerstone of cognitive security, its explainability and transparency are critical to build trust and comply with regulatory requirements. Cognitive security systems often operate as "black boxes," making decisions based on algorithms that humans cannot understand. Improved AI explainability entails creating models that provide human-friendly explanations of how decisions are made. For example, explainable AI might highlight the data points or patterns that triggered a security alert.

In many industries there are strict regulations that require organizations to demonstrate how decisions are made, especially when dealing with sensitive data. Improved AI explainability helps organizations meet laws like the GDPR through auditable and transparent decision-making processes. Transparent AI systems build trust among stakeholders such as employees, customers, and regulators. Cognitive security solutions can be more widely accepted if they can provide justifications for their actions.

13.6.3 Collaborative Threat Intelligence (CTI)

Cybersecurity is no longer an isolated endeavour; it calls for cooperation and shared intelligence to keep up with changing threats. CTI brings together insights from multiple organizations to provide a coordinated defence against cyberattacks. Industry-wide threat data sharing can help organizations better understand emerging attack vectors and techniques. As an example, industries like finance, health, and technology may benefit from shared data on malware signatures, phishing and ransomware attacks.

Advancements in cognitive security now allow sharing of threat intelligence in real time across secure platforms. This enables organizations to react faster to cyber threats and take preventive measures based on collective knowledge. Collaboration can benefit from data privacy and competition between organizations, however. Building trust and establishing secure frameworks for data sharing are critical to the success of CTI initiatives.

13.6.4 Autonomous Systems

The potential future of cognitive security consists of autonomous systems that can react to cyber threats without human intervention. Such systems can significantly reduce response times and improve the overall security system. Autonomous cognitive security systems detect, analyze, and neutralize threats in real time using AI and ML. As an example, a computer detecting a malware attack may isolate infected

For Product Safety Concerns and Information please contact our EU representative GPSR@taylorandfrancis.com Taylor & Francis Verlag GmbH, Kaufingerstraße 24, 80331 München, Germany

For Product Safety Concerns and Information please contact our EU representative GPSR@taylorandfrancis.com Taylor & Francis Verlag GmbH, Kaufingerstraße 24, 80331 München, Germany